Introduction to
Structural Equation Models

Introduction to Structural Equation Models

Otis Dudley Duncan

Department of Sociology
University of Arizona
Tucson, Arizona

ACADEMIC PRESS   New York San Francisco London

A Subsidiary of Harcourt Brace Jovanovich, Publishers

ACADEMIC PRESS, INC.
111 Fifth Avenue, New York, New York 10003

United Kingdom Edition published by
ACADEMIC PRESS, INC. (LONDON) LTD.
24/28 Oval Road, London NW1

Library of Congress Cataloging in Publication Data

Duncan, Otis Dudley.
    Introduction to structural equation models.

    (Studies in population)
    Bibliography: p.
    Includes index.
    1.   Sociology–Statistical methods.   2.   Sociology–
Mathematical models.   I.   Title.   II.   Series.
HM24.D93   301'.01'82   74-17988
ISBN 0–12–224150–9

Contents

Preface

The interests and level of preparation of graduate students and research workers in sociology were kept in mind in the writing of this volume. The material covered in the usual first-semester graduate course in statistics for sociologists (at least, I hope it is covered there) will suffice by way of preparation for its study. I have in mind the standard treatment of expectations of random variables, variances and covariances, regression (simple and multiple), and correlation. The notion of sets of linear equations and solutions thereof should have been acquired from college algebra. At a few points, determinants are used to facilitate the exposition, but matrices do not appear. Since this is not a textbook of statistics nor a manual of research methods, I have given no details on statistical and computational procedures.

The aim of the volume is to prepare the reader to understand the recent sociological literature reporting on the use of structural equation models in research and discussing methodological questions pertaining to such models. I believe the volume may also be appropriate for students of political science, but I could not have hoped in a short and elementary presentation to provide equally adequate orientation to the econometric and psychometric literature. Perhaps some students interested in these fields will find the discussion useful as a warm-up to the serious study of their subjects.

It is not my purpose to advocate or defend the use of structural equation models in sociology. Indeed, I hold a rather agnostic view about their ultimate utility and vitality in that discipline, fascinating as the models may be in purely formal terms. One thing that has given me pause is that I did not find it expedient to include here any substantive sociological examples, real or contrived. In thinking about the matter, it seemed to me that most of the persuasive real examples pertain to a narrowly delimited area of sociological inquiry. On the other hand, many contrived examples in the methodological literature of sociology are so utterly implausible as to call into question their very purpose. In econometrics, by contrast, there are well-known models—the "cobweb" model, for example, or a rudimentary Keynesian-type national income model—that are simple enough to serve as textbook examples but resemble serious models sufficiently to motivate their study.

The last thing I want to do is to suggest that the structural equation approach comprises a new recipe for conducting sociological research. I hope that no one, upon learning of this book, will write me for the computer program which I use to "do" structural equation models. I have no such program. What I am trying to say throughout the book is that "doing" the models consists largely in thinking about what kind of model one wants and can justify in the light of the ideas whose validity one is prepared to take responsibility for. Once the thinking has been done there is a role for computation. I, of course, believe in data analysis as a stimulus to thinking. But, as always, it is well to keep distinct the sources and the substance of one's ideas.

My original intention was to confine the work to the first 7 chapters. The material therein is almost entirely standard, so that upon its mastery the obvious next step is to begin studying a more advanced text. The remaining four chapters introduce progressively more open-ended issues. The purpose here is to seduce the reader into beginning to think for himself about the properties of models or even, on occasion, to suggest a problem that may intrigue the advanced student.

The first draft of this book was prepared as a report of the Institute for Advanced Studies and Scientific Research, Vienna, where I taught a course on structural equation models in the fall of 1973. I am grateful to the Institute for the opportunity to think about the subject in a systematic way for a sustained period of time and to test my thinking in the classroom. Astute questions put to me by the Scholaren (Michael Brenner, Peter Findl, Max Haller, Ingeborg Heinrich, Herbert Matschinger, Manfred Nermuth, and Friedrich Neumeyer) and the Abteilungsleiter (Jürgen M. Pelikan) of the Abteilung Soziologie provided the stimulus for my writing and required me to improve the presentation of many points.

The manuscript was read by Arthur S. Goldberger and Robert M. Hauser, both of whom noted a number of errors and suggested several improvements. I find that in this field, as in others, if you do not know your subject well it is a great help to have friends who do. In the light of my stubbornness on some matters and my prolificacy in generating errors, Goldberger and Hauser are not to be blamed for any of the book's faults, only for some of its virtues.

Beverly Duncan helped, too, in more than one way.

Introduction to
Structural Equation Models

1

Preliminaries

In this book we will frequently use a notation like the following:

$$u$$
$$\downarrow$$
$$x \longrightarrow y$$

It may be read as "a change in x or u produces a change in y" or "y depends on x and u" or "x and u are the causes of y." In all these statements, we need to include a distinction between two sorts of causes, x and u. We intend the former to stand for some definite, explicit factor (this variable has a name) producing variation in y, identified as such in our *model* of the dependence of y on its causes. On the other hand, u stands for all other sources (possibly including many different causes) of variation in y, which are not explicitly identified in the model. It sums up all their effects and serves to account for the fact that no single cause, x, nor even a finite set of causes (a list of specific x's) is likely to explain all the observable variation in y. (The variable u has no specific name; it is just called "the disturbance.")

The letters (like x, y, and u), the arrows, and the words (like "depends on") are elements in a language we use in trying to specify how we think the world—or, rather, that part of it we have selected for study—works. Once our ideas are sufficiently definite to help us make sense of the observations we have made or intend to make, it may be useful to formalize them in terms of a model. The preceding little arrow diagram, once we understand all the conventions for reading it, is actually a model or, if one prefers, a pictorial representation of a model. Such a representation has been found useful by many investigators as an aid in clarifying and conveying their ideas and in studying the properties of the models they want to entertain.

More broadly useful is the algebraic language of variables, constants, and functions, symbolized by letters and other notations, which are manipulated according to a highly developed grammar. In this language, our little model may be expressed as

$$y = bx + u$$

or, even more explicitly,

$$y = b_{yx} x + u$$

It is convenient, though by no means essential, to follow the rule that y, the "dependent variable" or "effect" is placed on the left-hand side of the equation while x, the "independent variable" or "cause," goes on the right-hand side. The constant, or coefficient b in the equation tells us by how much x influences y. More precisely, it says that a change of one unit in x (on whatever scale we adopt for the measurement of x) produces a change of b units in y (taking as given some scale on which we measure y). When we label b with subscripts (as in b_{yx}) the order of subscripts is significant: the first named variable (y) is the dependent variable, the second (x) the independent variable.

(**Warning:** *Although this convention will be followed throughout this book, not all authors employ subscripts to designate the dependent and independent variables.*)

The scale on which u is measured is understood to be the same as that used for y. No coefficient for u is required, for in one sense, u is merely a balancing term, the amount added to the quantity bx to satisfy the equation. (In causal terms, however, we think of y as depending on u, and not vice versa.) This statement may be clearer if we make explicit a feature of our grammar that has been left implicit up to now. The model is understood to apply (or, to be proposed for application) to the behavior of units in some population, and the variables y, x, and u are variable quantities or "measurements" that describe those units and their behavior. [The units may be individual persons in a population of people. But they could also be groups or collectives in a population of such entities. Or they could even be the occasions in a population of occasions as, for instance, a set of elections, each election being studied as a unit in terms of its outcome (y) and being characterized by properties such as the number of candidates on the ballot (x), for example.] We may make this explicit by supplying a subscript to serve as an identifier of the unit (like the numeral on the sweater of a football player). Then the equation of our model is

$$y_i = b_{yx}x_i + u_i$$

That is, for the ith member of the population we ascertain its score or value on x, to wit x_i, multiply it by b, and add to the product an amount u_i (positive or negative). The sum is equal to y_i, or the score of the ith unit on variable y. Ordinarily we will suppress the observation subscript in the interest of compactness, and the operation of summation, for example, will be understood to apply over all members of a sample of N units drawn from the population.

It is assumed that the reader will have encountered notation quite similar to the foregoing in studying the topic of regression in a statistics course. (Such study is a prerequisite to any serious use of this book.) But what we have been discussing is not statistics. Rather, we have been discussing the form of one kind of model that a scientist might propose to represent his ideas or theory about how things work in the real world. Theory construction, model building, and statistical inference are distinct activities, sufficiently so that there is strong pressure on a scientist to specialize in one of them to the exclusion of the others. We hope in this book to hint at reasons why such specialization should not be carried too far. But we must note immediately some reasons why the last two may come to be intimately associated.

Statistics, in one of its several meanings, is an application of the theory of probability. Whenever, in applied work—and all empirical inquiry is "applied" in this sense—we encounter a problem that probability theory may help to solve, we turn to statistics for guidance.

There are two broad kinds of problems that demand statistical treatment in connection with scientific use of a model like the one we are discussing. One is the problem of inference from samples. Often we do not have information about all units in a population. (The population may be hypothetically infinite, so that one could never know about "all" units; or for economic reasons we do not try to observe all units in a large finite population.) Any empirical estimate we may make of the coefficient(s) in our model will therefore be subject to sampling error. Any inference about the form of our model or the values of coefficients in it that we may wish to base on observational data will be subject to uncertainty. Statistical methods are needed to contrive optimal estimators and proper tests of hypotheses, and to indicate the degree of precision in our results or the size of the risk we are taking in drawing a particular conclusion from them.

The second, not unrelated, kind of problem that raises statistical issues is the supposition that some parts of the world (not excluding the behavior of scientists themselves, when making fallible measurements) may be realistically described as behaving in a stochastic (chance, probabilistic, random) manner. If we decide to build into our models some assumption of this kind, then we shall need the aid of statistics to formulate appropriate descriptions of the probability distributions.

This last point is especially relevant at this stage in the presentation of our little model, for there is one important stipulation about it that we have not yet stated. We think of the values of u as being drawn from a probability distribution. We said before that, for the ith unit of observation, u_i is the amount added to bx_i to produce y_i. Now we are saying that u_i itself is produced by a process that can be likened to that of drawing from a large set of well-mixed chips in a bowl, each chip bearing some value of u. The description of our model is not complete until we have presented the "specification on the disturbance term," calling u the "disturbance" in the equation (for reasons best known to the econometricians who devised the nomenclature), and meaning by the "specification" of the model a statement of the assumptions made about its mathematical form and the essential stochastic properties of its disturbance.

Throughout this book, we will assume that the values of the disturbance are drawn from the same probability distribution for all units in the population. This subsumes, in particular, the assumption of "homoskedasticity." It can easily be wrong in an empirical situation, and tests for departures from homoskedasticity are available. When the assumption is too wide of the mark, special methods (for example, transformation of variables, or weighting of regression estimators) are needed to replace the methods sketched in this book. No special attention is drawn to this assumption in the remainder of the text: but the reader must not forget it, nonetheless. Another assumption made throughout is that the mean value of the disturbance in the population is zero. The implications of assumptions about the disturbance are discussed in Chapter 11.

Frequently we shall assume explicitly that the disturbance is uncorrelated with the causal variable(s) in a model, although this assumption will be modified when the logic of the situation requires. Thus for the little model under study now, we specify that $E(xu) = 0$. (E is the sign for the expectation operator. If the reader is not familiar with its use in statistical arguments, he should look up the properties of the operator in an intermediate statistics text such as Hays, 1963, Appendix B.) The assumption that an explanatory or causal variable is uncorrelated with the disturbance must always be weighed carefully. It may be negated by the very logic of the model, as already hinted. If it is supposed, not only that y depends on x, but also that x simultaneously depends on y, it is contradictory to assume that the disturbance in the equation explaining y is uncorrelated with x. The specification $E(xu) = 0$ may also be contrary to fact, even when it is not inherently illogical. The difficulty is that we will never know enough about the facts of the case to be sure that the assumption is true—that would be tantamount to knowing everything about the causes of y. Lacking omniscience, we rely on theory to tell us if there are substantial reasons for faulting the assumption. If so, we shall have to eschew it—however convenient it may be—and consider how, if at all, we may modify our model or our observational procedures to remedy the difficulty. For *we must have this assumption in the model in some form*—though not necessarily in regard to all causal variables—*if any statistical procedures* (estimation, hypothesis testing) *are to be justified.* Here we distinguish sharply between (*1*) statistical description, involving summary measures of the joint distributions of observed variables, which may serve the useful purpose of data reduction, and (*2*) statistical methods applied to the problem of estimating coefficients in a *structural model* (as distinct from a "statistical model") and testing hypotheses about that model. One can do a passably good job of the former without knowing much about the subject matter (witness the large number of specialists in "multivariate data analysis" who have no particular interest in any substantive field). But one cannot even get started on the latter task without a firm grasp of the relevant scientific theory, because the starting point is, precisely, the model and not the statistical methods.

In summary, we have proposed a model,

$$y = b_{yx}x + u$$

and stated a specification on its disturbance term, $E(xu) = 0$. Without mentioning it before, we have also been assuming that $E(x) = 0$, which is simply a convention as to the location of the origin on the scale of the independent variable. It follows at once that

$$E(y) = b_{yx}E(x) + E(u) = 0.$$

Now, each of the variables in our model has a variance, and it is convenient to adopt the notation,

$$\sigma_{yy} = E(y^2)$$
$$\sigma_{xx} = E(x^2)$$
$$\sigma_{uu} = E(u^2)$$

for the variances (writing σ_{yy}, for example, in place of the usual σ_y^2). There are also three covariances,

$$\sigma_{yx} = E(yx)$$
$$\sigma_{yu} = E(yu)$$
$$\sigma_{xu} = E(xu) = 0$$

The disappearance of the last of these covariances is merely a restatement of the original specification on the disturbance term. To evaluate σ_{yu} we multiply the equation by u and take expectations, finding

$$E(yu) = b_{yx}E(xu) + E(uu)$$

so that $\sigma_{yu} = \sigma_{uu}$, in view of the fact that $E(xu) = 0$. Let us multiply through the equation of our model by y, obtaining,

$$y^2 = b_{yx}xy + yu$$

We take expectations

$$E(y^2) = b_{yx}E(xy) + E(yu)$$

and thereby find that we can write the variance of y as

$$\sigma_{yy} = b_{yx}\sigma_{xy} + \sigma_{uu}$$

since $\sigma_{yu} = \sigma_{uu}$ as already noted. Let us next multiply through by x and take expectations. We find

$$E(xy) = b_{yx}E(x^2) + E(xu)$$

or

$$\sigma_{xy} = b_{yx}\sigma_{xx}$$

Substituting this result into the expression for the variance of y, we obtain

$$\sigma_{yy} = b_{yx}^2\sigma_{xx} + \sigma_{uu}$$

(The same result is obtained upon squaring both sides of $y = b_{yx}x + u$ and taking expectations.)

The three symbols on the right-hand side stand for the basic parameters of this model as it applies in a well-defined *population*:

—the structural coefficient b_{yx},
—the variance of the exogenous variable x,
—the variance of the disturbance u.

The variance in the dependent variable is traceable to these three distinct sources.

The expression just obtained for the covariance of the two observable variables is a suggestive one, for we can immediately rewrite it as

$$b_{yx} = \frac{\sigma_{xy}}{\sigma_{xx}}$$

We see that if we knew σ_{xy} and σ_{xx} we could calculate the value of the structural coefficient. We do not and, in general, cannot know these quantities exactly. But we can estimate them, or their ratio, from data pertaining to a *sample* of the population to which the model applies. How to use this sample information in a correct and efficient manner is a topic studied in the statistical theory of estimation. In this book, we will draw upon a few important results from that theory, but will not try to demonstrate those results.

Exercise. *Our model*

$$y = bx + u$$

could be solved for x, to read

$$x = \frac{1}{b}y - \frac{1}{b}u$$

Let 1/b be renamed c and −u/b be called v. Someone could, therefore, assert that our model is equally well written

$$x = cy + v$$

But on the assumption that our original model is true (including the specification on the disturbance term) show that the disturbance is not uncorrelated with the variable on the right-hand side, that is, $E(yv) \neq 0$. Show also that we cannot solve for c using the same kind of formula developed for the original model, that is, $c \neq \sigma_{xy}/\sigma_{yy}$. How do you square this result with the well-known fact in statistics, that there are two regressions, Y on X and X on Y?

FURTHER READING

The statistical methods of simple and multiple regression are well presented in Snedecor and Cochran (1967, Chaps. 6, 7, and 13). A judicious discussion of issues raised in the use of the regression model to represent a causal relationship is given by Rao and Miller (1971, Chap. 1).

2

Correlation and Causation

Partly for historical reasons, the topic of structural equation models has often been approached by considering the implications of causal relationships for observable correlations or, inversely, the problem of rendering a causal interpretation of observed correlations. Taking the correlation coefficient as a point of departure has been particularly characteristic of psychometrics. But sociology, as well, has depended heavily on this measure of the degree of linear association between two quantitative variables. We suggest in Chapter 4 that a more fundamental view of structural equation models is secured by foregoing the algebra of correlation. Nevertheless, some aspects of our topic are quite convenient to develop in terms of correlation, so that is the way we shall begin.

Let us reconsider briefly the illustrative model studied in Chapter 1:

$$y = b_{yx}x + u$$

We already specified that $E(x) = E(u) = 0$ [whence it follows that $E(y) = 0$], and that $E(xu) = 0$. Let us now suppose that each variable (y, x, and u) is to be reexpressed in units of its own standard deviation. We can make the equality hold, while retaining the form of the model, by introducing ratios of standard deviations in the following fashion:

$$\frac{y}{\sigma_y} = b_{yx} \frac{\sigma_x}{\sigma_y} \cdot \frac{x}{\sigma_x} + \frac{\sigma_u}{\sigma_y} \cdot \frac{u}{\sigma_u}$$

The expressions

$$b_{yx} \frac{\sigma_x}{\sigma_y} = p_{yx}$$

$$\frac{\sigma_u}{\sigma_y} = p_{yu}$$

were termed *path coefficients* by Sewall Wright (1921, 1960, 1968), the great pioneer in the development of structural equation models (Li, 1956, 1968; Goldberger, 1972a). Let us now revise our notation, so that y, x, and u are the *standardized values* of the dependent variable, the independent variable, and the disturbance. Henceforth, in Chapters 2 and 3 only, we shall suppose that $E(y^2) = E(x^2) = E(u^2) = 1$, since the variance of a standardized variable is unity.

Writing our model in terms of path coefficients and standardized variables, we have

$$y = p_{yx} x + p_{yu} u$$

with the specification $E(xu) = 0$, which is not affected by standardization. (*Why?*) We may study the properties of this model, as before, by multiplying through the equation by one or another of the variables, taking expectations, and simplifying. The last step calls upon the theorem that the covariance of two standardized variables is the coefficient of correlation. Hence, using the letter *rho* to designate the correlation between two variables in the population under study (reserving the letter r for the corresponding correlation in a sample), we have

$$E(xy) = \rho_{xy}$$

We see at once that, since

$$E(xy) = p_{yx} E(x^2) + p_{yu} E(xu)$$

we have an equality of the path coefficient and the correlation,

$$\rho_{xy} = p_{yx}$$

recalling that $E(xu) = \rho_{xu} = 0$. (We hasten to add that the path coefficient has the same value as the correlation because this model is one with only a single explanatory variable; this will not hold true in general.) In writing path coefficients, as for structural coefficients, the first subscript refers to the variable affected, the second to the causal variable. In writing simple correlations, the order of subscripts is immaterial, since $\rho_{xy} = \rho_{yx}$.

The variance of the dependent variable is given by

$$E(y^2) = p_{yx}E(yx) + p_{yu}E(yu)$$

But, since $E(y^2) = 1$, we have

$$1 = p_{yx}\rho_{yx} + p_{yu}\rho_{yu}$$

Multiplying the equation of the model through by u, we find

$$E(yu) = p_{yx}E(xu) + p_{yu}E(uu)$$

so that

$$\rho_{yu} = p_{yu}$$

Hence, with one dependent variable, one explanatory variable, and the disturbance, we find that the variance (set at unity) of the dependent variable is partitioned into

$$1 = p_{yx}^2 + p_{yu}^2$$

the "explained" (by x) and "unexplained" portions.

As the reader must have guessed, this little model has been used primarily to illustrate notation, nomenclature, and the basic techniques to be used henceforth in studying properties of a model. It is time to complicate the discussion, and we do so by introducing a third explicit variable in addition to x and y. Three-variable models are interesting mainly for didactic purposes. But they illustrate all the essential principles that apply in more elaborate models.

Let us call the three variables x, y, and z and assume that each is in standard form. For the time being, we will suppose that nothing is known about the causal relationships—which (if any) of these variables cause(s) any of the others. It has sometimes been thought that correlations among variables can be employed in a quasi-deductive logic: If A and B are positively correlated, and if B and C are positively correlated, then it is quite likely that A and C are positively correlated. Whatever its plausibility in any particular concrete instance, this mode of reasoning clearly is not rigorous in general. If ρ_{xy}, ρ_{xz}, and ρ_{yz} are the correlations among three variables, then it can be shown that a certain determinant must be nonnegative; that is, that

$$\begin{vmatrix} 1 & \rho_{xy} & \rho_{xz} \\ \rho_{xy} & 1 & \rho_{yz} \\ \rho_{xz} & \rho_{yz} & 1 \end{vmatrix} = 1 + 2\rho_{xy}\rho_{yz}\rho_{xz} - \rho_{xy}^2 - \rho_{yz}^2 - \rho_{xz}^2 \geq 0$$

This theorem places only very broad constraints on the possible range of values of any one of the correlations. To see this, put in hypothetical values for two of the correlations, evaluate the determinant, and analyze the result. For example, if $\rho_{xz} = \rho_{yz} = .5$, we have

$$1 - 2\rho_{xy}^2 + \rho_{xy} \geq 0$$

so that ρ_{xy} can have any value between $-.5$ and $+1.0$. If $\rho_{xz} = \rho_{yz} = \sqrt{.5}$ (about .707), we have

$$\rho_{xy} - \rho_{xy}^2 \geq 0$$

so that ρ_{xy} is constrained to fall between zero and unity. Thus, one must assume rather high values for two of the correlations in order to decide with certainty even the sign of the third correlation.

If, on the other hand, we are in a position to assume causal relationships among the variables, rather strong deductions may be possible. To illustrate, consider the diagram

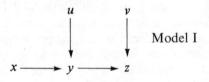

We interpret this as a representation of a two-equation model,

$$\left.\begin{aligned} y &= p_{yx}x + p_{yu}u \\ z &= p_{zy}y + p_{zv}v \end{aligned}\right\} \text{Model I}$$

where the p's are path coefficients and the specification on the disturbances is $\rho_{ux} = \rho_{vx} = \rho_{vy} = 0$. That is, each disturbance is uncorrelated with all "prior" causal variables. (But note that neither ρ_{uy},

ρ_{uz}, nor ρ_{vz} is zero.) Let us substitute the value of y, as given by the y-equation, into the z-equation:

$$z = p_{zy}p_{yx}x + p_{zy}p_{yu}u + p_{zv}v$$

Multiply through by x and take expectations:

$$\rho_{zx} = p_{zy}p_{yx}$$

since $\rho_{xu} = \rho_{xv} = 0$ and $E(x^2) = 1.0$. But both the y-equation and the z-equation are just like the model studied previously. So we already know that $p_{zy} = \rho_{zy}$ and $p_{yx} = \rho_{yx}$. Hence, we conclude that for this "simple causal chain" model

$$\rho_{zx} = \rho_{zy}\rho_{yx}$$

whatever the values of the two correlations on the right-hand side.

It is a plausible conjecture that when we attempt to reason from the values of two correlations to the value of a third, we must actually be working with an implicit causal model. With the right kind of causal model, the reasoning is valid. Without one, the reasoning is loose. Would it not be advantageous in such cases to make the causal model explicit, so as to be able to check our reasoning carefully?

In general, the advantages of having the model explicit are seen to lie in (1) making our arguments consistent (so that we are not altering our premises surreptitiously in the course of a discussion), (2) making our conclusions precise (so that it is easier to see what evidence is, and what is not, compatible with them), and thereby (3) rendering our conclusions susceptible of empirical refutation.

In the case of our causal chain, if $\rho_{zy} = .7$ and $\rho_{yx} = .6$, we infer that $\rho_{zx} = .42$. The conclusion is, indeed, precise. If it is belied by the facts, we must question one or more of the premises, for the argument itself is explicit and unexceptionable. There are two difficulties, one material, the other logical. The material difficulty is that we can never know ρ_{zx} exactly but can only estimate it from sample data. Hence the statement $\rho_{zx} \neq .42$ is based on a statistical inference and is, therefore, subject to uncertainty (the degree of uncertainty, however, we may hope to estimate by statistical methods). The logical difficulty is that, having rejected the conclusion of our argument because it is contradicted by the facts, we do not know which one(s) of the premises is (are) in error. We may have entered into the argument with erroneous values of either ρ_{zy} or ρ_{yx} or both. Moreover, we may have taken the wrong model as the basis for the argument. If, however, we are confident of the facts as to the two correlations and also reasonably confident that the conclusion is unsupported by evidence, we must reject the model, realizing the third advantage claimed for making the model explicit— albeit an advantage the full enjoyment of which calls for a certain taste for irony, if not masochism.

Consider another diagram:

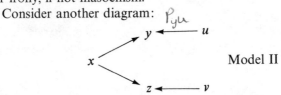

Model II

Again, we have two equations, but this time they have the same causal variable:

$$\left. \begin{array}{l} y = p_{yx}x + p_{yu}u \\ z = p_{zx}x + p_{zv}v \end{array} \right\} \text{Model II}$$

In each equation we specify a zero correlation between the causal variable and the disturbance, so that $\rho_{xu} = \rho_{xv} = 0$. We further specify a zero correlation between the disturbances of the two equations, $\rho_{uv} = 0$.

Exercise. *Show that in the case of Model I it was unnecessary to make this condition explicit, since it was implied by the specification, $\rho_{ux} = \rho_{vx} = \rho_{vy} = 0$.*

We now multiply through each equation by each variable in the model, take expectations, and express the result in terms of correlations.

Exercise. *Verify the following results.*

$$\rho_{yx} = p_{yx}$$

$$\rho_{zx} = p_{zx}$$

$$\rho_{yz} = p_{yx}p_{zx}$$

$$\rho_{uz} = \rho_{vy} = 0$$

$$\rho_{uy} = p_{yu}$$

$$\rho_{vz} = p_{zv}$$

We also find:

$$E(y^2) = 1 = p_{yx}^2 + p_{yu}^2$$
$$E(z^2) = 1 = p_{zx}^2 + p_{zv}^2$$

The key result for Model II is that $\rho_{yz} = \rho_{yx}\rho_{zx}$. Again we see the possibility of rejecting the model, upon discovering that the value of one of the correlations is inconsistent with the values of the other two, assuming Model II is true.

It is worth noting that this test of Model II is equivalent to a test of the null hypothesis that the partial correlation $\rho_{yz.x} = 0$. This is interesting because the tradition of empirical sociology has placed strong emphasis upon partial correlation, partial association, and "partialing" in general as a routine analytical procedure. By contrast, the approach taken in this book leaves rather little scope for partial correlation as such, although it does attempt to exploit the insights into causal relationships that often guide the capable analyst working with partial correlations.

To show the connection of partial correlation with Model II, recall the definition of a first-order partial correlation in terms of simple correlations:

$$\rho_{yz.x} = \frac{\rho_{yz} - \rho_{yx}\rho_{zx}}{\sqrt{1 - \rho_{yx}^2}\sqrt{1 - \rho_{zx}^2}}$$

We see at once that, if Model II holds, the numerator of this partial correlation is zero, since $\rho_{yz} = \rho_{yx}\rho_{zx}$, and thus $\rho_{yz.x} = 0$. Let us suppose that Model II has been rejected since $\rho_{yz.x}$ in the population under study clearly is not zero. We can represent the situation by relinquishing the specification $\rho_{uv} = 0$ which was originally part of the model. We require a new path diagram,

Model II′ $(\rho_{uv} \neq 0)$

in which the curved, double-headed arrow linking u and v stands for the correlation (ρ_{uv}) between these two variables (leaving unresolved the issue of whether that correlation has any simple causal interpretation). We note that the equations of Model II′ can be solved, respectively, for u and v,

$$u = \frac{y - p_{yx}x}{p_{yu}}$$

$$v = \frac{z - p_{zx}x}{p_{zv}}$$

from which it is but a short step to

$$E(uv) = \rho_{uv} = \frac{\rho_{yz} - \rho_{yx}\rho_{zx}}{p_{yu}p_{zv}}$$

in view of results already obtained. We now note that

$$p_{yu}^2 = 1 - p_{yx}^2 = 1 - \rho_{yx}^2$$

and

$$p_{zv}^2 = 1 - p_{zx}^2 = 1 - \rho_{zx}^2$$

so that we may write

$$\rho_{uv} = \frac{\rho_{yz} - \rho_{yx}\rho_{zx}}{\sqrt{1 - \rho_{yx}^2}\sqrt{1 - \rho_{zx}^2}}$$

But this is precisely the formula for $\rho_{yz.x}$ given earlier.

We see that Model II is a way of expressing the hypothesis that a single common cause (here called x) exactly accounts for the entirety of the correlation between two other variables (here, y and z). Model II′ asserts that y and z do indeed share x as a common cause, but that some other factor or factors, unrelated to x, also serve to induce a correlation between them. Note that the hypothesis underlying Model II is subject to a direct test, once we have sufficiently reliable estimates of the three observable correlations ρ_{yz}, ρ_{yx}, and ρ_{zx}. Model II′ is not so easily rejected. Indeed, one can (if one wishes) defend the truth of Model II′ in the face of any conceivable set of three correlations (that is, correlations that satisfy the condition stated for the determinant on page 12). To call Model II′ seriously into question really requires that we develop and justify a different model such that, if the new model is correct, Model II′ cannot be true. (When you think you have learned enough about the technique of working with structural equation models to try your hand at it, you should undertake the exercise of posing a countermodel to Model II', such that a set of observable correlations *could* show decisively that, if the new model is true, Model II' cannot be.)

We consider next a model in which one variable has the other two as causes:

$$z = p_{zy}y + p_{zx}x + v \qquad \text{Model III}$$

It takes only one equation to write this model. Since the equation has two variables that are causally prior to z, we specify that both are uncorrelated with the disturbance term: $\rho_{yv} = \rho_{xv} = 0$. The model as such says nothing about the correlation of y and x. In the absence of definite information to the contrary, we would do best to assume that this correlation may not be zero. Hence, the path diagram will be drawn as

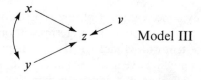

Model III

In some previous literature (for example, Blalock, 1962–1963), the contrary assumption led to Model III', as below, with $\rho_{xy} = 0$.

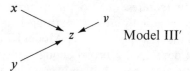

Model III'

Model III' has the ostensible advantage that one can test and possibly reject it. The "test" would consist merely in ascertaining whether a set of sample data are consistent with the hypothesis, $\rho_{xy} = 0$. But this "test" does not get at the *causal* properties of the model in any way. Indeed, one should think of the value of ρ_{xy} as being entirely exogenous in this model. The model might well hold good in different populations with varying values (not excluding zero) of ρ_{xy}. We will have to work harder to get a cogent test of Model III.

Let us multiply through the equation of Model III by each of the causal variables, take expectations, and express the result in terms of correlations. (By now, this technique should be quite familiar.) We find

$$\left.\begin{array}{l} \rho_{yz} = p_{zy} + p_{zx}\rho_{xy} \\ \rho_{xz} = p_{zy}\rho_{xy} + p_{zx} \end{array}\right\} \quad \text{(from Model III)}$$

This set of equations may be looked at from two viewpoints. First, it shows how the correlations involving the dependent variable are generated by the path coefficients and the exogenous correlation, ρ_{xy}. Second, if we know the values of the three correlations in a population to which Model III applies, we may solve these equations uniquely for the path coefficients:

$$p_{zy} = \frac{\rho_{zy} - \rho_{zx}\rho_{xy}}{1 - \rho_{xy}^2}$$

$$p_{zx} = \frac{\rho_{zx} - \rho_{zy}\rho_{xy}}{1 - \rho_{xy}^2}$$

We note that any conceivable set of three correlations (that is, correlations meeting the condition on the determinant on page 12) will be consistent with the truth of Model III. (It may happen, incidentally, that a path coefficient will have a value greater than unity or less than -1.0, for this coefficient is not constrained in the same way that a correlation is.) Thus to call Model III into question—other than by questioning the plausibility of estimated values of the path coefficients—really requires one to pit it against some alternative model, supported by equally strong or stronger theoretical considerations, such that if the new model is true Model III is necessarily faulty. (We have not yet developed a broad enough grasp of the possibilities to exemplify this strategy here; but upon completing your study of this book, you should be able to employ it with confidence.)

We have now surveyed the whole field of possibilities for the class of three-variable models involving just two causal paths and no "feedback" or "reciprocal" relationships. Other models of the kinds already studied can be developed, however, merely by interchanging the positions of the variables in Model I, II, or III. It is instructive to study the entire set of models that can be formulated in this way.

In the diagrams in Table 2.1 the disturbances are not given letter names (as they would be in the algebraic representation of these models), but are symbolized by the arrows that originate outside the

Table 2.1   Three-Variable Recursive Models with 2 Causal Arrows

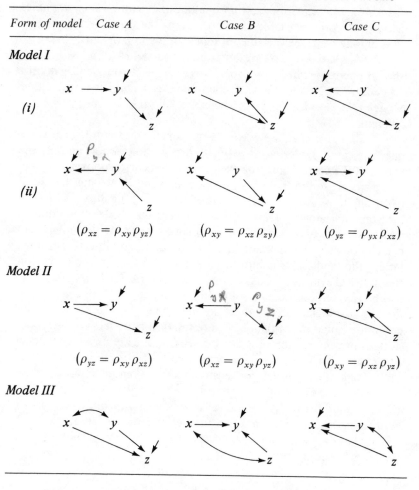

Form of model	Case A	Case B	Case C

diagram. The specifications on the disturbances are the same, in form, as those already stated for Models I, II, and III.

Exercise.   *Derive the conditions written below Models IB, IC, IIB, and IIC. For each of Models IIIB and IIIC, derive a pair of equations expressing correlations in terms of path coefficients.*

With the aid of this figure, it is worth pondering the problem of "causal inference." Suppose that one does not know which variables cause which other variables, but does know the values of the correlations among the three variables in some population. Imagine that these correlations satisfy the condition given for Model IA(i) $(\rho_{xz} = \rho_{xy}\rho_{yz})$. Should one report his "discovery" that z depends on y and y in turn depends on x? No, for this same condition is also consistent with either Model IA(ii) or Model IIB, where the causal relationship is quite different.

You should immediately leap to the conclusion that one can *never* infer the causal ordering of two or more variables knowing only the values of the correlations (or even the partial correlations!).

We can reason in the other direction, however. Knowing the causal ordering, or, more precisely, the causal model linking the variables, we can sometimes infer something about the correlations. Or, assuming a model for sake of argument, we can express its properties in terms of correlations and (sometimes) find one or more conditions that must hold if the model is true but that are potentially subject to refutation by empirical evidence. Thus any one of Models IA, IB, IC, IIA, IIB, IIC can be rejected on the basis of reliable estimates of the three correlations, if their estimated values do not come close to satisfying the condition noted in the figure. Failure to reject a model, however, does not require one to accept it, for the reason, already noted, that some other model(s) will always be consistent with the same set of data. Only if one's theory is comprehensive and robust enough to rule out all these alternatives would the inference be justified.

We turn more briefly to three-variable models with one causal arrow connecting each pair of variables in one direction or the other.

There are actually only two kinds of models. One is the recursive model of the form

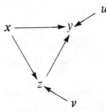

in which each variable depends on all "prior" or "predetermined" variables.

Exercise.   *Enumerate the six possible models of this general form, interchanging variables.*

The three-variable recursive model is included within the four-variable recursive model treated in the next chapter, so we shall not investigate it further here.

The second kind of model involving three causal arrows is represented by either of the diagrams below:

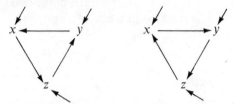

As we shall discover in the next chapter, neither of these is a recursive model. For reasons that are probably not clear now, but should become so upon study of nonrecursive models, we are no longer entitled to assume that the disturbance in a particular equation is uncorrelated with the causal variable in that equation. Indeed, that assumption is not merely contrary to fact, it gives rise to logical contradictions. The kinds of deductions made for Models I, II, and III are no longer legitimate. Moreover, with nonrecursive models it is often not plausible to assume that the disturbances in the several equations are uncorrelated among themselves (although this assumption can be made if there are adequate grounds for doing so). In that event, we should confront a model like the following:

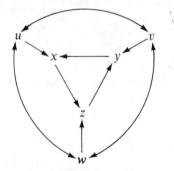

But this model is "underidentified," the meaning and implications of which condition we shall consider in Chapter 6.

Note on Partial Correlation

In the foregoing we paid no attention to partial correlations except when comparing Model II with Model II', in which the partial correlation $\rho_{yz.x} = \rho_{uv}$ is actually a parameter of the model. This omission may seem odd to readers acquainted with the literature on "causal inference" in sociology, wherein partial correlations seem to play a central role. It is true, of course, that each of the conditions tabulated on page 19 could be restated in terms of a certain partial correlation. Thus, if, as for Model IA,

$$\rho_{xz} = \rho_{xy}\rho_{yz}$$

it will also be true that

$$\rho_{xz.y} = \frac{\rho_{xz} - \rho_{xy}\rho_{yz}}{\sqrt{1 - \rho_{xy}^2}\sqrt{1 - \rho_{yz}^2}} = 0.$$

Accordingly, we might treat the sample partial correlation $r_{xz.y}$ as a test statistic for the null hypothesis, $\rho_{xz} = \rho_{xy}\rho_{yz}$. But we shall find in Chapter 3 that it is not necessary to invoke the concept of partial correlation in order to carry out an equivalent test.

We also note that the partial correlation, unless it has a well-defined role as a parameter of the model or as a test statistic, may actually be misleading. Continuing with the example of Model IA, suppose it occurred to an investigator to estimate the partial correlation $\rho_{zy.x}$—not for any clearly defined purpose, but simply because it is a "good idea" to look at the partial correlations when one is working with three or more variables. We know that, by definition,

$$\rho_{zy.x} = \frac{\rho_{zy} - \rho_{zx}\rho_{yx}}{\sqrt{1 - \rho_{zx}^2}\sqrt{1 - \rho_{yx}^2}}$$

If Model IA is true, we have seen that $\rho_{zx} = \rho_{zy}\rho_{yx}$, so we may substitute that value into the foregoing formula to see what $\rho_{zy.x}$ would be under this model:

$$\rho_{zy.x} = \frac{\rho_{zy} - \rho_{zy}\rho_{yx}^2}{\sqrt{1 - \rho_{zx}^2}\sqrt{1 - \rho_{yx}^2}}$$

$$= \rho_{zy}\sqrt{\frac{1 - \rho_{yx}^2}{1 - \rho_{zx}^2}}$$

Now, if all the correlations are less than unity and $\rho_{zx} = \rho_{zy}\rho_{yx}$ (as it does when the model is true), we must have $\rho_{zx}^2 < \rho_{zy}^2$ and also $\rho_{zx}^2 < \rho_{yx}^2$. It follows that the fraction on the right-hand side is less than unity, so that

$$|\rho_{zy.x}| < |\rho_{zy}|$$

or, if ρ_{zy} is positive,

$$\rho_{zy.x} < \rho_{zy}$$

Now, it is not always clear what one does after "partialing," but it seems likely that the investigator who goes so far as to estimate $\rho_{zy.x}$ will not leave the result unreported. Most likely he will announce something like, "The relationship between z and y is reduced when x is held constant," and perhaps offer reasons why this might happen. But, if Model IA is true, this result is artifactual, and the investigator's report is erroneous.

We might also note, in light of our work with Model III, that if the investigator had computed, not the partial correlation $\rho_{zy.x}$, but the corresponding partial *regression* coefficient, the outcome would not be misleading. For the value of that regression coefficient is

$$\frac{\rho_{zy} - \rho_{zx}\rho_{yx}}{1 - \rho_{yx}^2}$$

or, upon substituting $\rho_{zy}\rho_{yx}$ for ρ_{zx} on the assumption that Model I is true,

$$\frac{\rho_{zy} - \rho_{zy}\rho_{yx}^2}{1 - \rho_{yx}^2} = \rho_{zy} = p_{zy}$$

That is, the partial regression coefficient is the same as the path coefficient, as it should be for this kind of model, and the numerical result is intelligible in terms of the model, while the one for the partial correlation is misleading.

We find, therefore, that partial correlations may only be a distraction when studying some of the models treated in this chapter (see also Duncan, 1970). Indeed the very concept of correlation itself has a subordinate role in the development of structural equation models. Although in Chapter 3 the presentation of recursive models does feature correlations, in subsequent chapters both recursive and nonrecursive models are treated from a different and more fundamental point of view.

FURTHER READING

The classic exposition of the causal structures that may underlie the correlations among three variables is that of Simon (1954). A didactic presentation of four-variable models much in the spirit of this chapter is given by Blalock (1962–1963). Both papers are reprinted in Blalock (1971).

3

Recursive Models

A model is said to be recursive if all the causal linkages run "one way," that is, if no two variables are reciprocally related in such a way that each affects and depends on the other, and no variable "feeds back" upon itself through any indirect concatenation of causal linkages, however circuitous. However, recursive models do cover the case in which the "same" variable occurs at two distinct points in time, for, in that event, we would regard the two measurements as defining two different variables. For example, a dynamic model like the following:

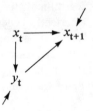

where t and $t + 1$ are two points in time, is recursive (even though x appears to feed back upon itself). The definition also subsumes the case in which there are two or more ostensibly contemporaneous dependent variables where none of them has a direct or indirect causal linkage to any other. This situation is illustrated by Models II and II' in Chapter 2 (assuming that ρ_{uv} in Model II' does not implicitly arise from either a path $y \rightarrow z$ or a path $z \rightarrow y$). In this case, we have to consider whether or not to specify $\rho_{uv} = 0$. If so, we might term the model "fully recursive"; if not, it is merely "recursive."

With the exception of this last kind of situation—which offers no difficulty in principle, though it requires careful handling in practice— we can state that all the *dependent variables* in a recursive model (those whose causes are explicitly represented in the model) are arrayed in an unambiguous causal ordering. Moreover, all *exogenous variables* (those whose causes are not explicitly represented in the model) are, as a set, causally prior to all the dependent variables. There is, however, no causal ordering of the exogenous variables (if there are two or more) with respect to each other. (In some other model, of course, these variables might be treated as dependent variables.)

It simplifies matters greatly and results in a more powerful model if we can assume there is only one exogenous variable. This may not always be a reasonable assumption. We will consider models of this kind first and then see what modifications are entailed if we have to assume the contrary.

All our exposition of recursive models will rest on illustrations in which there are just four variables. Throughout this book we rely on relatively simple examples, and it is expected that the reader will come to see how the principles pertaining to these examples can be generalized to other models. There is some risk in this procedure, but it is hopefully outweighed by the advantages of making the discussion both concrete and compact—virtues difficult to attain if all principles and theorems must be stated in a perfectly general form.

With four variables, one of them exogenous, the causal ordering will involve a unique arrangement. Any one of the four might be the exogenous variable, any of the remaining three might be the first dependent variable, and either of the remaining two might follow it. Hence, there are $4 \cdot 3 \cdot 2 \cdot 1 = 24$ possible causal orderings of four variables. To build a recursive model means that we must choose one and only one of these 24 as the "true" ordering. That choice, it should be clear from the discussion in the previous chapter, cannot be based on the correlations among the variables, because *any* correlation matrix will be consistent with *any* causal ordering one may propose.

The information in regard to causal ordering is, logically, a priori. Such information is derived from theory, broadly construed, and no amount of study of the formal properties of models can teach one how to come up with a true theory. We can, however, prescribe the task of theory—to provide a causal ordering of the variables. Another way to put it is that the theory must tell us that at least six of the twelve possible causal linkages among four variables are *not* present, and these missing links, moreover, must fall into a triangular pattern. For, if four variables are put into a causal order and then numbered in sequence, we will have a pattern like the following:

		Caused by		
Effect	x_1	x_2	x_3	x_4
(x_1)	$\cdots$	0	0	0
x_2	$\times$	$\cdots$	0	0
x_3	$\times$	$\times$	$\cdots$	0
x_4	$\times$	$\times$	$\times$	$\cdots$

It is immaterial whether one or more of the crosses is replaced by a nought. But all the noughts must be present, to stand for the assumption that x_2 does *not* cause x_1 (though x_1 may cause x_2), and so on. If one enters six noughts in such a matrix (ignoring diagonal cells) before the variables are numbered, it may or may not be possible to triangulate the matrix. If it is, then the matrix defines a recursive causal ordering. If not, one or more nonrecursive relationships is present. Thus, the indispensable contribution of theory is to put noughts into the matrix. It is an odd way to put it, but the decisive criterion of the utility of a theory is that it can tell us definitely what causal relationships do not obtain, not that it can suggest (however evocatively) what relationships may well be present. (This remark will seem even more poignant when, in a later chapter, we discuss the problem of identification for a nonrecursive model.)

There is still another way to describe a recursive model. We may say that all exogenous variables in the model are *predetermined* with respect to all dependent variables. Moreover, each dependent variable is predetermined with respect to any other dependent variable that occurs later in a causal ordering. We take this to mean that all exogenous variables are uncorrelated with the disturbances in all equations of the model. (If one cannot assume this, the remedy is to build a better model.) Moreover, we shall similarly assume that the predetermined variables occurring in any equation are uncorrelated with the disturbance of that equation. (This is virtually a definition of "predetermined." It does leave open the possibility that of two dependent variables in a recursive model, neither is predetermined with respect to the other, while the disturbances of their equations are correlated, as is the case for Model II′ in the previous chapter.) Again, this assumption has to be evaluated on its theoretical or substantive merits and, if it must be faulted, the recourse is to propose a better model.

The four-variable model, already described by a matrix of crosses and noughts, is more explicitly represented by this set of equations:

$$(x_1 \text{ exogenous})$$

$$x_2 = p_{21}x_1 + p_{2u}x_u$$

$$x_3 = p_{32}x_2 + p_{31}x_1 + p_{3v}x_v$$

$$x_4 = p_{43}x_3 + p_{42}x_2 + p_{41}x_1 + p_{4w}x_w$$

or by the path diagram:

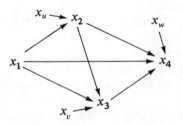

We continue to assume that $E(x_h) = 0$ and $E(x_h^2) = 1$, $h = 1,2,3,4$, u,v,w (the variables are in standard form). Hence, $E(x_h x_j) = \rho_{hj}$, the correlation (in the population) between x_h and x_j. The specifications on the disturbance terms are as follows:

(*a*) The exogenous variable is uncorrelated with the disturbances:

$$E(x_1 x_u) = E(x_1 x_v) = E(x_1 x_w) = 0$$

(b) Disturbances are also uncorrelated with any other predetermined variables in an equation:

$$E(x_2 x_v) = E(x_2 x_w) = E(x_3 x_w) = 0$$

A standard manipulation is to "multiply through" one equation of the model by a variable in the model, take expected values, and express in terms of path coefficients (the p's) and correlations (ρ's). Following this procedure with the x_2-equation, making use of (a) and (b), we deduce that the disturbance in each equation is uncorrelated with the disturbance in any other equation; for example,

$$E(x_2 x_v) = p_{21} E(x_1 x_v) + p_{2u} E(x_u x_v)$$

so that $\rho_{uv} = 0$; similarly, $\rho_{uw} = \rho_{vw} = 0$. But note that the disturbance in each equation has a nonzero correlation with the dependent variable in that equation and (in general) with the dependent variable in each "later" equation. To take another example, if we multiply through the equation for x_3 by x_2, we obtain

$$E(x_2 x_3) = p_{32} E(x_2^2) + p_{31} E(x_1 x_2) + p_{3v} E(x_2 x_v)$$

or

$$\rho_{23} = p_{32} + p_{31} \rho_{12}$$

[since $E(x_2^2) = 1$ and $E(x_2 x_v) = 0$]. Proceeding systematically, we obtain

$$\rho_{12} = p_{21} \qquad \text{(from the } x_2\text{-equation)}$$

$$\left. \begin{aligned} \rho_{13} &= p_{31} + p_{32} \rho_{12} \\ \rho_{23} &= p_{31} \rho_{12} + p_{32} \end{aligned} \right\} \quad \text{(from the } x_3\text{-equation)}$$

$$\left. \begin{aligned} \rho_{14} &= p_{41} + p_{42} \rho_{12} + p_{43} \rho_{13} \\ \rho_{24} &= p_{41} \rho_{12} + p_{42} + p_{43} \rho_{23} \\ \rho_{34} &= p_{41} \rho_{13} + p_{42} \rho_{23} + p_{43} \end{aligned} \right\} \quad \text{(from the } x_4\text{-equation)}$$

We may study this set of "normal equations" in two ways:

(i) *Solve for the p's in terms of the ρ's*

We obtain, for the first equation of the model,

$$p_{21} = \rho_{12}$$

for the second equation,

$$p_{31} = (\rho_{13} - \rho_{12}\rho_{23})/(1 - \rho_{12}^2)$$
$$p_{32} = (\rho_{23} - \rho_{12}\rho_{13})/(1 - \rho_{12}^2)$$

and for the third equation,

$$p_{41} = \frac{1}{D} \begin{vmatrix} \rho_{14} & \rho_{12} & \rho_{13} \\ \rho_{24} & 1 & \rho_{23} \\ \rho_{34} & \rho_{23} & 1 \end{vmatrix}$$

$$p_{42} = \frac{1}{D} \begin{vmatrix} 1 & \rho_{14} & \rho_{13} \\ \rho_{12} & \rho_{24} & \rho_{23} \\ \rho_{13} & \rho_{34} & 1 \end{vmatrix}$$

$$p_{43} = \frac{1}{D} \begin{vmatrix} 1 & \rho_{12} & \rho_{14} \\ \rho_{12} & 1 & \rho_{24} \\ \rho_{13} & \rho_{23} & \rho_{34} \end{vmatrix}$$

where

$$D = \begin{vmatrix} 1 & \rho_{12} & \rho_{13} \\ \rho_{12} & 1 & \rho_{23} \\ \rho_{13} & \rho_{23} & 1 \end{vmatrix}$$

Thus, if we knew the correlations we could solve for the coefficients of the model. In practice, we have only estimates (from a sample) of the correlations. If these estimates are inserted into the foregoing formulas (in place of the population correlations), the formulas will yield estimates of the p's. These are, in fact, the same estimates that one obtains from the ordinary least squares (OLS) regression of

$$x_2 \text{ on } x_1$$

$$x_3 \text{ on } x_2 \text{ and } x_1$$

$$x_4 \text{ on } x_3, x_2, \text{ and } x_1$$

if the variables are in standard form.

(ii) Solve for the ρ's in terms of the p's

This may be done quite simply, making substitutions in the "normal equations." We obtain

$$\rho_{12} = p_{21}$$

$$\rho_{13} = p_{31} + p_{32}p_{21}$$

$$\rho_{23} = p_{32} + p_{31}p_{21}$$

$$\rho_{14} = p_{41} + p_{42}p_{21} + p_{43}(p_{31} + p_{32}p_{21})$$

$$\rho_{24} = p_{42} + p_{43}p_{32} + p_{41}p_{21} + p_{43}p_{31}p_{21}$$

$$\rho_{34} = p_{43} + p_{42}(p_{32} + p_{31}p_{21}) + p_{41}(p_{31} + p_{32}p_{21})$$

These expressions are instructive in that they show how the model generates the (observable) correlations. It is worthwhile to study them with some care.

(1) We see that the entirety of the correlation between x_1 and x_2 is generated by the direct effect, p_{21}

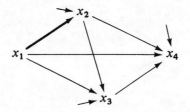

(2) The correlation between x_1 and x_3 is generated by two distinct paths, so that ρ_{13} equals the direct effect, p_{31}

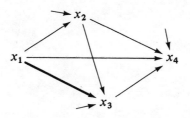

plus the indirect effect, $p_{32} p_{21}$

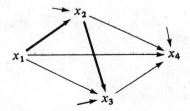

(3) The situation is different in regard to x_2 and x_3, for here we have the total correlation (ρ_{23}) generated as the sum of the direct effect, p_{32}

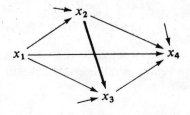

plus correlation due to a common cause, $p_{31} p_{21}$

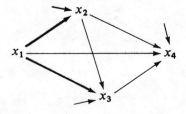

(4) The correlation between x_1 and x_4 is generated by four distinct causal links; ρ_{14} equals the direct effect, p_{41}

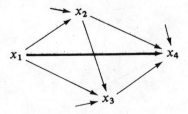

plus indirect effect via x_2, $p_{42}p_{21}$

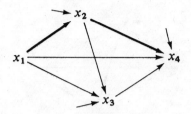

plus indirect effect via x_3, $p_{43}p_{31}$

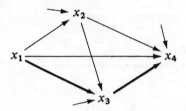

plus indirect effect via x_3 and x_2, $p_{43}p_{32}p_{21}$

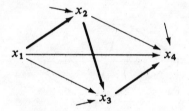

(5) Both an indirect effect and correlation due to common causes are involved in generating the correlation between x_2 and x_4; ρ_{24} equals the direct effect, p_{42}

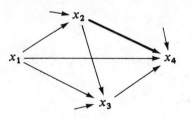

plus indirect effect via x_3, $p_{43} p_{32}$

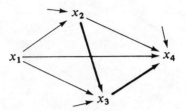

plus correlation due to x_1 operating as a common cause, directly, $p_{41} p_{21}$

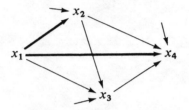

and indirectly (via x_3), $p_{43} p_{31} p_{21}$

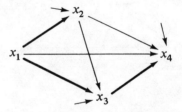

(6) There are no indirect effects in the model producing correlation between x_3 and x_4; but there are two common causes. Hence, ρ_{34} equals the direct effect, p_{43}

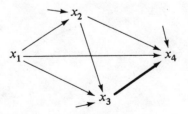

plus correlation due to common causes, working directly, $p_{42}p_{32}$

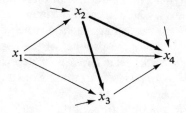

and $p_{41}p_{31}$

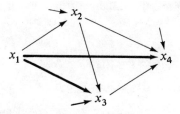

or indirectly, $p_{42}p_{31}p_{21}$

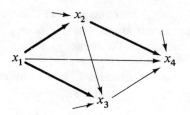

and $p_{41}p_{32}p_{21}$

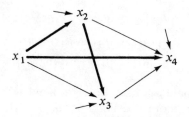

In all six of these cases, the correlation may be read off the path diagram using Wright's (1921) multiplication rule: To find the correlation between x_h and x_j, where x_j appears "later" in the model, begin at x_j and read *back* to x_h along each distinct direct and indirect (compound) path, forming the product of the coefficients along that path. After reading back, read *forward* (if necessary), but only one reversal from back to forward is permitted. Sum the products obtained for all the linkages between x_j and x_h.

We consider next a modification of the model. Suppose both x_1 and x_2 are exogenous, that is, the model cannot explain how they are generated. Since we do not know anything about this, we cannot make any strong assumption about their correlation. Hence, we shall have to suppose, in general, that $\rho_{12} \neq 0$. The model now has but two equations,

$$x_3 = p_{32}x_2 + p_{31}x_1 + p_{3v}x_v$$

$$x_4 = p_{43}x_3 + p_{42}x_2 + p_{41}x_1 + p_{4w}x_w$$

Disturbances are uncorrelated with predetermined (including exogenous) variables. Hence, $\rho_{1v} = \rho_{1w} = \rho_{2v} = \rho_{2w} = \rho_{3w} = 0$; and, as a consequence, $\rho_{vw} = 0$. Normal equations are obtained as before, except, of course, that there is no normal equation with ρ_{12} on the left-hand side. Also, except for the fact that no p_{21} occurs in the model, the solution for the p's is the same as before and has the same interpretation.

In studying the model from standpoint *ii* (solution for ρ's in terms of p's), however, we must reconsider the situation. Algebraic substitutions yield,

$$\rho_{13} = p_{31} + p_{32}\rho_{12}$$

$$\rho_{23} = p_{32} + p_{31}\rho_{12}$$

$$\rho_{14} = p_{41} + p_{43}p_{31} + (p_{42} + p_{43}p_{32})\rho_{12}$$

$$\rho_{24} = p_{42} + p_{43}p_{32} + (p_{41} + p_{43}p_{31})\rho_{12}$$

$$\rho_{34} = p_{43} + p_{42}p_{32} + p_{41}p_{31} + (p_{42}p_{31} + p_{41}p_{32})\rho_{12}$$

We cannot eliminate ρ_{12} from the right-hand side of these equations, since the model cannot (by definition) tell us anything about how that correlation is generated. Moreover, the presence of this correlation means that the remaining correlations are generated in a somewhat ambiguous way. Consider the correlation between x_1 and x_3. We have a direct effect, p_{31}. The other term, $p_{32}\rho_{12}$, consists of the product of the direct effect of x_2 on x_3 and the correlation of x_1 and x_2. It represents a contribution to ρ_{13} by virtue of the fact that *another cause* of x_3 (namely x_2) is correlated (to the extent of ρ_{12}) with the cause we are examining at the moment (namely, x_1). As in Chapter 2 (see Models II′ and III) we use a curved, double-headed arrow to refer to a correlation that cannot be analyzed in terms of causal components within this model. Hence the way to look at this situation is that ρ_{13} equals the direct effect, p_{31}

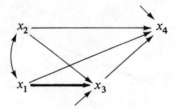

plus correlation due to correlation with another cause, $p_{32}\rho_{12}$

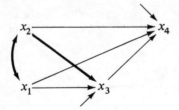

Similarly, we may break down ρ_{23} into the components: direct effect, p_{32}

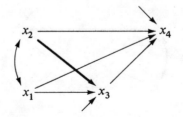

plus correlation due to correlation with another cause, $p_{31}\rho_{12}$

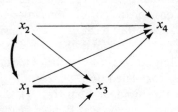

The same kind of reasoning interprets the remaining correlations. For ρ_{14} we obtain the direct effect, p_{41}

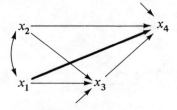

plus the indirect effect, $p_{43}p_{31}$

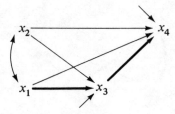

plus correlation due to the correlation of x_1 with another cause (x_2), working both directly, $p_{42}\rho_{12}$

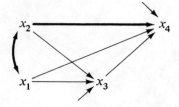

and indirectly, $p_{43}p_{32}\rho_{12}$

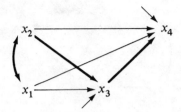

We decompose ρ_{24} into the direct effect, p_{42}

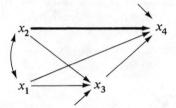

plus the indirect effect, $p_{43}p_{32}$

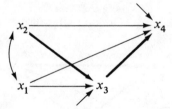

plus correlation due to the correlation of x_2 with another cause (x_1), working both directly, $p_{41}\rho_{12}$

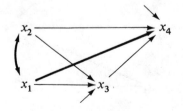

and indirectly, $p_{43} p_{31} \rho_{12}$

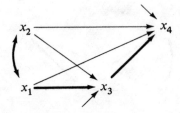

As in the previous version of the model, ρ_{34} involves no indirect effects, but the correlation generated by the common causes (x_1 and x_2) involves both the direct effects of those causes and the correlation due to the fact that they are correlated with each other. Hence, ρ_{34} equals the direct effect, p_{43}

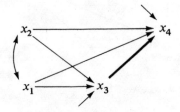

plus correlation due to x_2 as a common cause, $p_{42} p_{32}$

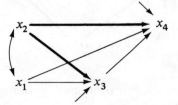

and x_1 as a common cause, $p_{41} p_{31}$

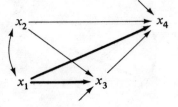

plus correlation due to the correlation of x_1 with another common cause (x_2), $p_{42} p_{31} \rho_{12}$

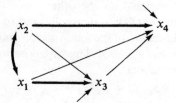

and correlation due to the correlation of x_2 with another common cause (x_1), $p_{41} p_{32} \rho_{12}$

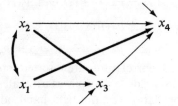

For some purposes, one might be content to aggregate the last four components, so as to describe the correlation ρ_{34} as being generated by the direct effect (p_{43}) and the correlation due to the influence of the two common causes, x_1 and x_2, on x_3 and x_4.

To bring this discussion within the scope of Sewall Wright's multiplication rule, we stipulate that the curved double-headed arrow is read either back or forward.

It is, of course, an undesirable property of this modified model that we cannot clearly disentangle the effects of its two exogenous variables. But our theory may simply be unable to tell us whether x_1 causes x_2, x_2 causes x_1, each influences the other, both are effects of one or more common or correlated causes, or some combination of these situations holds true. In that event, we cannot know for sure whether a change initiated in (say) x_1 will have indirect effects via x_2 or not, since we do not know whether x_2 depends on x_1. It follows that we cannot, with this model, estimate the *total* effect (defined as *direct* effect plus *indirect* effect) of x_1 on, say, x_4. There may or may not be an indirect causal linkage from x_1 through x_2 to x_4. But since we know nothing about this, we cannot include any such indirect effect in our estimate of total effect.

It should be noted, in both forms of the model, that the zero-order correlation between two variables often is not the correct measure of total effect of one variable on the other, since that correlation may include components other than direct and indirect effects.

In a further modification of the four-variable model, we suppose that x_1, x_2, and x_3 all are exogenous. This gives rise to the degenerate case of a single-equation model:

$$x_4 = p_{43}x_3 + p_{42}x_2 + p_{41}x_1 + p_{4w}x_w$$

The disturbance is uncorrelated with the exogenous variables.

There are three normal equations:

$$\rho_{14} = p_{41} + p_{42}\rho_{12} + p_{43}\rho_{13}$$

$$\rho_{24} = p_{41}\rho_{12} + p_{42} + p_{43}\rho_{23}$$

$$\rho_{34} = p_{41}\rho_{13} + p_{42}\rho_{23} + p_{43}$$

As before, if sample correlations are inserted into these equations, the solution for the p's yields least-squares estimates of the parameters.

None of the correlations on the right-hand side of the normal equations can be expressed in terms of path coefficients. Therefore, we cannot separate indirect effects from correlation due to common or correlated causes. Thus, the only decomposition we can provide is the following:

Causal variable	Total correlation	=	Direct effect on x_4	+	Correlation due to common and/or correlated causes
x_1	ρ_{14}		p_{41}		$p_{42}\rho_{12} + p_{43}\rho_{13}$
x_2	ρ_{24}		p_{42}		$p_{41}\rho_{12} + p_{43}\rho_{23}$
x_3	ρ_{34}		p_{43}		$p_{41}\rho_{13} + p_{42}\rho_{23}$

The path diagram for this single-equation model is shown below:

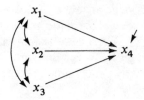

Correlations between exogenous variables are represented by curved, double-headed arrows. The normal equations can be written using Sewall Wright's rule. The curved arrow can be read either forward or backward, but *only one* curved arrow can be included in a given trajectory. Thus, to find ρ_{14} (for example), we read p_{41}

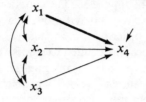

plus $p_{42}\rho_{12}$

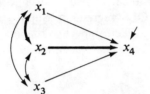

plus $p_{43}\rho_{13}$

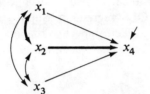

The single-equation model correctly represents the *direct effects* of the exogenous variables. But, since the causal structure of relationships among the exogenous variables is unknown, this model cannot tell us anything about how the indirect effects (not to mention total effects) are generated. In this respect, it is even less satisfactory than the two-equation model.

A major goal of theory, therefore, should be to supply a model that will make some of the exogenous variables endogenous.

This discussion has illustrated a significant theorem: The *direct* effects of predetermined variables in one equation of a model on the dependent variable of that equation are the same, irrespective of the causal relationships holding among the predetermined variables. Thus, returning to the complete model discussed at the beginning, the values of p_{41}, p_{42}, and p_{43} in the third equation do not depend on whether x_2 causes x_1 or vice versa in the first equation or on whether x_1, x_2, or x_3 is the dependent variable in the second equation.

Thus far we have only considered models in which all direct paths allowed by the causal ordering are, in fact, present in the model. We must now consider procedures suited to recursive models in which one or more such paths are (or may be) missing. These procedures have to do with the distinct (though related) problems of *estimation* and *testing*. We discuss them in that order.

We have seen that in the fully recursive model with direct paths from each "earlier" variable to each "later" variable, the path coefficients may be estimated by OLS regression. Suppose, however, that our model, while recursive, explicitly specifies that one or more coefficients are zero. To take a concrete example, consider this path diagram

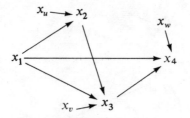

The equations of the model are

$$(x_1 \text{ exogenous})$$

$$x_2 = p_{21}x_1 + p_{2u}x_u$$

$$x_3 = p_{32}x_2 + p_{31}x_1 + p_{3v}x_v$$

$$x_4 = p_{43}x_3 + p_{41}x_1 + p_{4w}x_w$$

The only change from the model on page 28 is that $p_{42} = 0$. We continue to assume that all variables (including disturbances) are in standard form. In each equation of the model, the disturbance is uncorrelated with the predetermined variables. Moreover, we take the model to be fully recursive, so that the disturbance in each equation is uncorrelated with predetermined variables in all "earlier" equations. (Here, as elsewhere in this chapter, we allow the zero correlation of two disturbances to appear as a consequence of the zero correlation of disturbances with prior predetermined variables.) The force of this specification, with special reference to the present example, is that $\rho_{2w} = 0$, even though x_2 does not appear (explicitly) in the x_4-equation. We have, then, the following specification on the disturbances: $\rho_{1u} = \rho_{1v} = \rho_{1w} = \rho_{2v} = \rho_{2w} = \rho_{3w} = 0$. As a consequence of this specification, we find that it is also true that $\rho_{uv} = \rho_{uw} = \rho_{vw} = 0$.

The normal equations for the x_2-equation and the x_3-equation are the same as before, and OLS estimates of their path coefficients are obtained by formulas given earlier.

Multiplying through the x_4-equation by each predetermined variable, we find

$$\rho_{14} = p_{41} + p_{43}\rho_{13}$$

$$\rho_{24} = p_{41}\rho_{12} + p_{43}\rho_{23}$$

$$\rho_{34} = p_{41}\rho_{13} + p_{43}$$

Assuming the ρ's are known, we have three equations in the two unknown path coefficients. In mathematical terms, the solution for the p's is overdetermined. In the language of structural equation models, the x_4-equation is *overidentified*. In the event that an equation in the model is overidentified, we may deduce that one or more *overidentifying restrictions* must hold if the model is true. Here, we can ascertain the overidentifying restriction by writing out each of the solutions for the p's obtained upon solving a pair of the normal equations. There are three distinct solutions. If the model holds, the values obtained in all three must be equal. Thus,

$$\text{(i)} \qquad\qquad \text{(ii)} \qquad\qquad \text{(iii)}$$

$$p_{41} = \frac{\rho_{14} - \rho_{13}\rho_{34}}{1 - \rho_{13}^2} = \frac{\rho_{14}\rho_{23} - \rho_{13}\rho_{24}}{\rho_{23} - \rho_{12}\rho_{13}} = \frac{\rho_{24} - \rho_{23}\rho_{34}}{\rho_{12} - \rho_{13}\rho_{23}}$$

$$p_{43} = \frac{\rho_{34} - \rho_{13}\rho_{14}}{1 - \rho_{13}^2} = \frac{\rho_{24} - \rho_{12}\rho_{14}}{\rho_{23} - \rho_{12}\rho_{13}} = \frac{\rho_{12}\rho_{34} - \rho_{13}\rho_{24}}{\rho_{12} - \rho_{13}\rho_{23}}$$

where solution number (i) makes use of the first and third normal equations, number (ii) is from the first and second normal equations, and number (*iii*) is from the last two normal equations. If $p_{41}^{(i)} = p_{41}^{(ii)}$ it follows that

$$\rho_{24} + \rho_{13}\rho_{14}\rho_{23} + \rho_{12}\rho_{13}\rho_{34} - \rho_{24}\rho_{13}^2 - \rho_{23}\rho_{34} - \rho_{12}\rho_{14} = 0$$

We note that this expression is just the expansion of the determinant given as the numerator of p_{42} on page 30; and that determinant must be zero if $p_{42} = 0$. We reach the same conclusion from any of the other equalities of solutions for p_{41} or p_{43}. These several equalities are not independent. There is actually only one overidentifying restriction on this model.

Now, the overidentifying restriction must hold in any *population* in which the model applies. But if we have only *sample* values of the correlations we cannot expect it to hold exactly, nor can we expect the three solutions for each path coefficient to be exactly equal. In that event, to estimate the path coefficients we must choose one of the solutions, or perhaps some average of them. Since each solution makes use of only two of the normal equations, it would appear that averaging the solutions would be advisable, since we would then be making use of all the sample correlations rather than only some of them. *This intuition, however, is wrong.* It turns out that the preferred estimate is obtained upon inserting sample correlations into solution (*i*). It will be noted that the estimates of p_{41} and p_{43} obtained in this way are just the OLS regression coefficients of x_4 on x_1 and x_3. The general rule, then, is this: In a fully recursive model (where the correlation between each pair of disturbances is zero), estimate the coefficients in each equation by OLS regression of the dependent variable on the predetermined variables included in that equation.

The basis for this rule is a proof that the sampling variance of a $\hat{p}$ estimated by OLS is smaller than the variance of any other unbiased estimate of the same coefficient, even if such an estimate appears to use more information in the sense of combining correlations involving the included variables with correlations involving the excluded predetermined variable(s). Some of the earlier literature on path analysis was in error on this point; it was called to the attention of sociologists by A. S. Goldberger (1970).

We now turn to the problem of *testing*. In the preceding example, we discussed estimation on the assumption that the model and, in particular, the overidentifying restriction on the model are known in advance to be true. But the investigator may not feel confident of this specification. Indeed, he may be undertaking a study precisely to test that aspect of his theory which says that a particular coefficient should be zero. Much of the literature on causal models in the 1960s—particularly papers discussing or using the so-called "Simon–Blalock technique"—focused on this very question. Sometimes the problem was described as that of "making causal inferences from correlational data," but that ambiguous phrase seems to promise far too much. In the light of our preceding discussion, it would be more accurate to describe the problem as that of *testing the overidentifying restriction(s)* of a model.

We first take note of two plausible and conceptually correct procedures for making such tests on recursive models. But, since these procedures are not convenient from the standpoint of the standard methods of statistical inference, we conclude with an alternative recommendation.

Continuing with the example already described, suppose the analyst computes estimates of the path coefficients, p_{41} and p_{43}, by some method (not necessarily the OLS estimates recommended above). If these estimates—call them $\tilde{p}_{41}$ and $\tilde{p}_{43}$—are combined with sample correlations according to the normal equations, we have

$$r_{14}^* = \tilde{p}_{41} + \tilde{p}_{43} r_{13}$$

$$r_{24}^* = \tilde{p}_{41} r_{12} + \tilde{p}_{43} r_{23}$$

$$r_{34}^* = \tilde{p}_{41} r_{13} + \tilde{p}_{43}$$

where r_{hj} is an observed sample correlation and r_{hj}^* is the "implied" ("predicted" or "reproduced") correlation that would be observed if the overidentifying restriction(s) held exactly in the sample. Because of sampling error, implied and observed correlations will ordinarily not all be equal. Thus, we have a set of discrepancies,

$$d_{14} = r_{14} - r_{14}^*$$

$$d_{24} = r_{24} - r_{24}^*$$

$$d_{34} = r_{34} - r_{34}^*$$

If the OLS method of estimation were used, we would have

$d_{14} = d_{34} = 0$ but $d_{24} \neq 0$. If some other method were used we would still find one or more d's differing from zero. However, if the model holds true in the population, any such difference(s) should be "small," that is, no larger than one might reasonably expect as a consequence of sampling error alone.

It would seem plausible to use the set of d's in a formal statistical test against the null hypothesis which asserts the truth of the overidentifying restriction(s) of the model. However, this procedure is less convenient than the standard test described later, in the event that this test is available. Under some circumstances—though not in the case of the model used as an example here—the method of implied correlations may be recommended cautiously as a heuristic expedient, if an appropriate standard statistical test is not available. It may also be of use in the initial stage of specifying a model, where the investigator wishes to make an informal test of his ideas.

A similar approach to testing of overidentifying restrictions is the Simon–Blalock procedure. Consider any fully recursive model in which one or more paths are taken to be missing, that is, to have the value zero. It is then possible to deduce that certain simple and/or partial correlations will be zero. Blalock (1962–1963) has actually provided an exhaustive enumeration of the "predictions" for all possible four-variable models. The example given here appears in Blalock's enumeration as "Model E," and we find that in this model $\rho_{24.13} = 0$. The corresponding sample partial correlation ($r_{24.13}$) should, therefore, be close to zero. If it is not—if the difference from zero is too great to attribute to sampling error—we should be obliged to call into question the overidentifying restriction of this model. Blalock does not develop formal procedures of statistical inference for this kind of test. Again, we conclude that the proposed test is conceptually valid and is useful to the investigator who wants to be sure he understands the properties of his model. The research worker should not, however, rely on mere inspection of sample partial correlations. We recommend instead the standard statistical test described hereafter.

In our example, the issue as to the specification of the model is whether $p_{42} = 0$ or $p_{42} \neq 0$. In other words, we must decide as between the competing specifications of the x_4-equation:

$$x_4 = p_{43} x_3 + p_{41} x_1 + p_{4w} x_w$$

and

$$x_4 = p_{43}x_3 + p_{42}x_2 + p_{41}x_1 + p_{4w}x_w$$

We proceed on the latter specification and estimate by OLS the equation which includes p_{42}. In the usual routine for multiple regression we obtain as a by-product of our calculations the quantities necessary to compute the standard errors of our estimated coefficients (see, for example, the chapter on multiple regression in Walker and Lev, 1953, where the procedures are described for standardized variables). We may then form the ratio,

$$t = \hat{p}_{42}/\text{S.E.}(\hat{p}_{42})$$

and refer it to the t-distribution with the appropriate degrees of freedom. Roughly speaking, if one is working with a reasonably "large" sample, when $|t| \geq 2.0$, we may conclude with no more than 5% risk of error that the null hypothesis is false. In this event, we would reject the overidentifying restriction of the model and, presumably, respecify it to include a nonzero value of p_{42}.

In the case of failure to reject the null hypothesis—that is, if the t-ratio is not statistically significant—the situation is intrinsically ambiguous. Clearly, one is not obliged to *accept* the null hypothesis unless there is sufficient a priori reason to do so. It could happen, for example, that the true value of p_{42} is positive but small, so that our sample is just not large enough to detect the effect reliably. If our theory strongly suggests this is the case, we would do well to keep p_{42} in the equation despite the outcome of the test. In any event, it is good practice to publish standard errors of all coefficients, so that the reader of the research report may draw his own conclusion as well as have some idea of the precision of the estimates of coefficients. A good discussion of the issues raised by tests of this kind is given by Rao and Miller (1971); their discussion is presented in the context of a single-equation model, but it carries over to the problem of testing an overidentifying restriction on any equation of a recursive model.

This is not the place to develop the theory and techniques of statistical inference. What has been said can be reduced to a simple rule: If OLS regression is the appropriate method of estimation, the theory and techniques of statistical inference, as presented in the literature on the multiple regression model, should be drawn upon when making tests of overidentifying restrictions. (Perhaps we should note that such a test will not always involve the t-statistic for a single coefficient but may lead to the F-test for a whole set of coefficients.) This rule does not cover all cases, as will become apparent when we consider models for which OLS is not the appropriate method of estimation.

It is vital to keep the matter of tests of overidentifying restrictions in perspective. Valuable as such tests may be, they do not really bear upon what may be the most problematical issue in the specification of a recursive model, that is, the causal ordering of the variables. It is the gravest kind of fallacy to suppose that, from a number of competing models involving different causal orderings, one can select the true model by finding the one that comes closest to satisfying a particular test of overidentifying restrictions. (Examples of such a gross misunderstanding of the Simon–Blalock technique can be found, among other places, in the political science literature of the mid-1960s.) In fact, a test of the causal ordering of variables is beyond the capacity of any statistical method; or, in the words of Sir Ronald Fisher (1946), "if . . . we choose a group of social phenomena with no antecedent knowledge of the causation or absence of causation among them, then the calculation of correlation coefficients, total or partial, will not advance us a step toward evaluating the importance of the causes at work [p. 191]."

Exercise. *Show how to express all the correlations in terms of path coefficients in the model on page 44, using Sewall Wright's multiplication rule for reading a path diagram.*

Exercise. *Change the model on page 28 so that* p_{43} *is specified to be zero. Draw the revised path diagram. Obtain an expression for the overidentifying restriction. Indicate how one would estimate the coefficients in the revised model, supposing the overidentifying restriction was not called seriously into question.*

FURTHER READING

See Walker and Lev (1953, Chapter 13) for a treatment of multiple regression with standardized variables. Sociological studies using recursive models are numerous; for examples, see Blalock (1971, Chapter 7) and Duncan, Featherman, and Duncan (1972).

4

Structural Coefficients
in Recursive Models

In Chapter 3 a four-variable recursive model was formulated in terms of standardized variables. That procedure has some advantages.

(1) Certain algebraic steps are simplified.
(2) Sewall Wright's rule for expressing correlations in terms of path coefficients can be applied without modification.
(3) Continuity is maintained with the earlier literature on path analysis and causal models in sociology.
(4) It shows how an investigator whose data are available only in the form of a correlation matrix can, nevertheless, make use of a clearly specified model in interpreting those correlations.

Despite these advantages (see also Wright, 1960), it would probably be salutary if research workers relinquished the habit of expressing variables in standard form. The main reason for this recommendation is that standardization tends to obscure the distinction between the structural coefficients of the model and the several variances and covariances that describe the joint distribution of the variables in a certain population.

Although it will involve some repetition of ideas, we will present the four-variable recursive model again, this time avoiding the stipulation that all variables have unit variance. The model is

$$(x_1 \text{ exogenous})$$

$$x_2 = b_{21}x_1 + u$$

$$x_3 = b_{32}x_2 + b_{31}x_1 + v$$

$$x_4 = b_{43}x_3 + b_{42}x_2 + b_{41}x_1 + w$$

We continue to assume that all variables have zero expectation; unlike standardization, this achieves a useful simplification without significant loss of generality or confusion of issues. (Only in special cases, where "regression through the origin" is involved, does this stipulation require modification.)

The specification on the disturbances is that the disturbance in each equation has zero covariance with the predetermined variables in that equation and all "earlier" equations. The one strictly exogenous variable x_1 is predetermined in each equation. In addition, x_2 is a predetermined variable in the x_3-equation and the x_4-equation, while x_3 is a predetermined variable in the x_4-equation. Thus we specify $E(x_1 u) = E(x_1 v) = E(x_1 w) = E(x_2 v) = E(x_2 w) = E(x_3 w) = 0$. We find, as a consequence of this specification, that it also is the case that $E(uv) = E(uw) = E(vw) = 0$.

(We note, for future reference, that in both recursive and nonrecursive models, the usual specification is zero covariance of predetermined variables in an equation with the disturbance of that equation. In the case of recursive models this generally implies zero covariances among the disturbances of different equations. This does not, however, hold true for nonrecursive models.)

To deduce properties of the model from the specifications regarding its functional form and its disturbances, we multiply through equations of the model by variables in the model, and take expectations. For convenience, we denote the variance $E(x_j^2)$ by σ_{jj}, using the double subscript in place of the exponent of the usual notation, σ_j^2. Similarly, a covariance is denoted by $\sigma_{hj} = E(x_h x_j)$, $h \neq j$.

The normal equations are obtained by multiplying through each equation by its predetermined variables:

$$\sigma_{12} = b_{21}\sigma_{11} \qquad\qquad \text{(from the } x_2\text{-equation)}$$

$$\left.\begin{array}{l} \sigma_{13} = b_{31}\sigma_{11} + b_{32}\sigma_{12} \\[4pt] \sigma_{23} = b_{31}\sigma_{12} + b_{32}\sigma_{22} \end{array}\right\} \qquad \text{(from the } x_3\text{-equation)}$$

$$\left.\begin{array}{l} \sigma_{14} = b_{41}\sigma_{11} + b_{42}\sigma_{12} + b_{43}\sigma_{13} \\[4pt] \sigma_{24} = b_{41}\sigma_{12} + b_{42}\sigma_{22} + b_{43}\sigma_{23} \\[4pt] \sigma_{34} = b_{41}\sigma_{13} + b_{42}\sigma_{23} + b_{43}\sigma_{33} \end{array}\right\} \qquad \text{(from the } x_4\text{-equation)}$$

Clearly, it is possible to solve uniquely for the b's, which we shall term the _structural coefficients_, in terms of the population variances and covariances. In practice, of course, the latter are unknown. Hence, we can only estimate the structural coefficients. If, in the normal equations, the σ's are replaced by sample moments, the estimates obtained are equivalent to those of ordinary least-squares (OLS) regression of x_2 on x_1; x_3 on x_2 and x_1; and x_4 on x_3, x_2, and x_1. By sample moments we mean the quantities

$$m_{jj} = \sum x_j^2$$
$$m_{hj} = \sum x_h x_j \qquad (h \neq j)$$

where the summation is over all the observations in the sample and each observation on x_j is expressed as a deviation from the mean of x_j in the sample.

Along with the preceding normal equations, we will find it useful to obtain expressions for the variances of the dependent variables by multiplying through each equation of the model by its dependent variable:

$$\sigma_{22} = b_{21}\sigma_{12} + \sigma_{2u}$$
$$\sigma_{33} = b_{32}\sigma_{23} + b_{31}\sigma_{13} + \sigma_{3v}$$
$$\sigma_{44} = b_{43}\sigma_{34} + b_{42}\sigma_{24} + b_{41}\sigma_{14} + \sigma_{4w}$$

These may be simplified slightly by noting that $\sigma_{2u} = \sigma_{uu}$, $\sigma_{3v} = \sigma_{vv}$, and $\sigma_{4w} = \sigma_{ww}$, which facts are deduced by multiplying through each equation of the model by its disturbance.

It is instructive to rewrite the expressions for the variances and covariances in the form given below (the algebra involves only straightforward, though tedious, substitutions in the equations already given). The variances may be written:

$$\sigma_{11} \text{ exogenous}$$

$$\sigma_{22} = b_{21}^2 \sigma_{11} + \sigma_{uu}$$

$$\sigma_{33} = A_6 \sigma_{11} + b_{32}^2 \sigma_{uu} + \sigma_{vv}$$

$$\sigma_{44} = A_9 \sigma_{11} + A_0 \sigma_{uu} + b_{43}^2 \sigma_{vv} + \sigma_{ww}$$

and the covariances:

$$\sigma_{12} = b_{21} \sigma_{11}$$

$$\sigma_{13} = A_1 \sigma_{11}$$

$$\sigma_{23} = A_3 \sigma_{11} + b_{32} \sigma_{uu}$$

$$\sigma_{14} = A_2 \sigma_{11}$$

$$\sigma_{24} = A_4 \sigma_{11} + A_5 \sigma_{uu}$$

$$\sigma_{34} = A_7 \sigma_{11} + A_8 \sigma_{uu} + b_{43} \sigma_{vv}$$

where

$$A_1 = b_{31} + b_{32} b_{21}$$

$$A_2 = b_{41} + b_{42} b_{21} + b_{43} A_1$$

$$A_3 = b_{21}(b_{31} + b_{32} b_{21})$$

$$A_4 = b_{41} b_{21} + b_{42} b_{21}^2 + b_{43} A_3$$

$$A_5 = b_{42} + b_{43} b_{32}$$

$$A_6 = b_{32} A_3 + b_{31} A_1$$

$$A_7 = b_{41} A_1 + b_{42} A_3 + b_{43} A_6$$

$$A_8 = b_{32}(b_{42} + b_{43} b_{32})$$

$$A_9 = b_{43} A_7 + b_{42} A_4 + b_{41} A_2$$

$$A_0 = b_{42} A_5 + b_{43} A_8$$

The A's have been introduced merely to abbreviate the presentation and have no particular interpretation in themselves. It is important to note, however, that all the A's are nonlinear combinations of the structural coefficients (the b's) and *involve no other terms*. Thus, we can draw an important conclusion. The variances and covariances are all functions of (at most) three kinds of quantities: (*1*) the variance of the exogenous variable; (*2*) the variance(s) of one or more disturbances; and (*3*) a nonlinear combination of structural coefficients. Table 4.1

Table 4.1   Sources of Observable Variances and Covariances

Variance or covariance	Is a function of									
	σ_{11}	σ_{uu}	σ_{vv}	σ_{ww}	b_{21}	b_{31}	b_{32}	b_{41}	b_{42}	b_{43}
σ_{11}	×	...	...	...	...	...	...	...	...	...
σ_{22}	×	×	...	...	×	...	...	...	...	...
σ_{33}	×	×	×	...	×	×	×	...	...	...
σ_{44}	×	×	×	×	×	×	×	×	×	×
σ_{12}	×	...	...	...	×	...	...	...	...	...
σ_{13}	×	...	...	...	×	×	×	...	...	...
σ_{14}	×	...	...	...	×	×	×	×	×	×
σ_{23}	×	×	...	...	×	×	×	...	...	...
σ_{24}	×	×	...	...	×	×	×	×	×	×
σ_{34}	×	×	×	...	×	×	×	×	×	×

makes this explicit in each instance. The first component (σ_{11}) is involved in all the variances and covariances. One or more of the disturbance variances $(\sigma_{uu}, \sigma_{vv}, \sigma_{ww})$ are involved in the variances of all the dependent variables in the model and in the covariances of these variables with each other. Some combination of structural coefficients (the b's) is involved in all variances (except σ_{11}) and covariances. Thus, it is possible to regard the variances and covariances of the observed variables as having arisen entirely from these three sources. Moreover, we can describe these components as separable, in the following sense. We can suppose without contradiction (that is, without violating any other property of the model) that one of these components may change without either of the others having to change. If any of them changes, however, the observable variances and covariances will, in general, change.

This remarkable property of the model should be considered carefully by the investigator, for it has some far-reaching implications.

Suppose we had two populations under study and we specified our three-equation model as holding in each. It could happen that the structural coefficients are the same in the two populations and the variances of the several disturbances are likewise the same. But if only σ_{11} differs between the two populations, we will observe differences in all the other variances and all the covariances. Incidentally, we will also observe differences (in general) in all the correlations in the two populations and also (in general) in all the standardized path coefficients, not to mention other purported measures of "relative importance" or "unique contribution" of variables. Thus the observable facts about the two populations (as reflected in sample estimates of variances, covariances, and correlations) will suggest that they differ in many ways. But the premise of the illustration is that they differ in only one way: with respect to the variance of the exogenous variable. The model is invariant across populations with respect to the structural coefficients and the variances of the disturbances.

Another possibility is that both σ_{11} and the disturbance variances differ as between two populations, but the structural coefficients are the same. Again we would observe entirely different variance-covariance (or correlation) matrices in the two populations, even though only four of the ten quantities in the column headings of the table actually differ.

The possibilities just described are only hypothetical. But there would not be much purpose in devising a model to use in interpreting data if we did not have some hope that at least some features of our model would be invariant with respect to some changes in the circumstances under which it is applied. If all the model is good for is to describe a particular set of data—if with any new set of data we will be obliged to change all the quantities listed in the column headings of the table, even though we continue to specify the same mathematical form of the model—then we might as well forego the effort of devising and estimating the model. It offers no economy of description, since there are as many parameters across the top of the table (neglecting the possibility that some b's may be zero and therefore may be omitted) as there are variances and covariances listed in the stub. We can transform the variances and covariances into the parameters, or vice versa, by mathematical operations already described in these pages. Hence, from a purely descriptive standpoint, we might as well let stand the first set of estimates we compute—the variances and covariances, or correlations—and not bother with the structural coefficients.

Another line of reflection is suggested by this analysis. It could happen that a macrosociological model proposed for a given society gave rise to (nearly) invariant estimates of structural coefficients over a period of time (for example, for several successive birth cohorts), even though the variances of the exogenous variable and the disturbances were changing. One would then observe "social change" with respect to variances and covariances of the dependent variables. But in another sense, no "social change" would be occurring, since the structural coefficients were staying constant. If, on the other hand, the latter should change, one would really be dealing with social change, in a deeper sense of the term. The first kind of social change would, in a sense, be "explained" by the model (though, of course, the model does not speak to the sources of change in the exogenous variable and disturbances themselves). The second kind of social change—modification of structural coefficients (or "structural change," if one likes)—cannot be explained in any sense by the model. Even so, one might argue, the model—if one held to it with good reason, despite changes in structural coefficients—would at least make clear what it is about the social changes occurring that requires explanation. But surely our scientific aspirations and efforts should be directed toward the construction of models which are themselves "explanatory" in a proper scientific sense of the word, and not merely in the sense of providing some parametrization of the descriptive statistics which serves merely as a clue to the task of scientific explanation.

We gain still another perspective on the concept of structural coefficients in learning how to transform the model into a different set of equations. We continue with the model presented at the beginning of this chapter. By straightforward substitution we eliminate x_2 from the x_3-equation and both x_2 and x_3 from the x_4-equation; the x_2-equation is repeated as it stands. These manipulations yield the following three equations as the *reduced form* of the model:

(x_1 exogenous)

$$x_2 = b_{21}x_1 + u$$

$$x_3 = (b_{31} + b_{32}b_{21})x_1 + b_{32}u + v$$

$$x_4 = [b_{41} + b_{42}b_{21} + b_{43}(b_{31} + b_{32}b_{21})]x_1 + (b_{42} + b_{43}b_{32})u$$
$$+ b_{43}v + w$$

To obtain a compact notation for the coefficients and disturbances of the reduced form, we rewrite the foregoing equations, making use of the following definitions:

$$a_{21} = b_{21}$$

$$a_{31} = b_{31} + b_{32}b_{21}$$

$$a_{41} = b_{41} + b_{42}b_{21} + b_{43}(b_{31} + b_{32}b_{21})$$

$$u' = u$$

$$v' = b_{32}u + v$$

$$w' = (b_{42} + b_{43}b_{32})u + b_{43}v + w$$

This yields

$$x_2 = a_{21}x_1 + u'$$

$$x_3 = a_{31}x_1 + v'$$

$$x_4 = a_{41}x_1 + w'$$

We find that the exogenous variable is uncorrelated with the reduced-form disturbances, since

$$E(x_1u') = E(x_1u) = 0$$

$$E(x_1v') = b_{32}E(x_1u) + E(x_1v) = 0$$

$$E(x_1w') = (b_{42} + b_{43}b_{32})E(x_1u) + b_{43}E(x_1v) + E(x_1w) = 0$$

making use of the initial specification on the disturbances in the model. However, in the reduced form (unlike the *structural form* in which the model was originally specified), it is no longer true that disturbances are uncorrelated among themselves. In fact, we can derive explicit expressions for the covariances among reduced-form disturbances (recalling that the covariances among structural-form disturbances are zero):

$$\sigma_{u'v'} = E(u'v') = b_{32}E(u^2) + E(uv) = b_{32}\sigma_{uu}$$

$$\sigma_{u'w'} = (b_{42} + b_{43}b_{32})\sigma_{uu}$$

$$\sigma_{v'w'} = b_{32}(b_{42} + b_{43}b_{32})\sigma_{uu} + b_{43}\sigma_{vv}$$

The reduced-form disturbance variances likewise are functions of structural coefficients and variances of structural-form disturbances:

$$\sigma_{u'u'} = \sigma_{uu}$$

$$\sigma_{v'v'} = b_{32}^2\sigma_{uu} + \sigma_{vv}$$

$$\sigma_{w'w'} = (b_{42} + b_{43}b_{32})^2\sigma_{uu} + b_{43}^2\sigma_{vv} + \sigma_{ww}$$

Suppose we regard the expressions for the variances and covariances of the reduced-form disturbances, taking these quantities as known, as six equations in the six unknowns, σ_{uu}, σ_{vv}, σ_{ww}, b_{43}, b_{42}, and b_{32}. Although the equations involve nonlinear combinations of the unknowns, they are readily solved by a sequence of simple substitutions. Having solved for these parameters, we could return to the three equations defining the a's (taking them as known) and solve for the remaining unknown structural coefficients, b_{41}, b_{31}, and b_{21}.

Thus, if the reduced-form coefficients (the a's) and the variance–covariance matrix of the reduced-form disturbances were known, we could solve for structural coefficients (the b's) and the variances of structural disturbances. From a computational point of view, this result is of no great practical value. Because of its conceptual interest, however, we indicate how one might proceed. Multiplying each reduced-form equation through by the exogenous variable and taking expectations, we obtain:

$$\sigma_{12} = a_{21}\sigma_{11}$$

$$\sigma_{13} = a_{31}\sigma_{11}$$

$$\sigma_{14} = a_{41}\sigma_{11}$$

Least squares estimates of the a's are, therefore,

$$\hat{a}_{21} = \frac{m_{12}}{m_{11}}$$

$$\hat{a}_{31} = \frac{m_{13}}{m_{11}}$$

$$\hat{a}_{41} = \frac{m_{14}}{m_{11}}$$

Multiplying through each reduced-form equation by each dependent variable and each reduced-form disturbance yields (after a little algebraic manipulation, which may serve as an exercise for the interested reader):

$$\sigma_{u'u'} = \sigma_{22} - a_{21}\sigma_{12}$$

$$\sigma_{v'v'} = \sigma_{33} - a_{31}\sigma_{13}$$

$$\sigma_{w'w'} = \sigma_{44} - a_{41}\sigma_{14}$$

$$\sigma_{u'v'} = \sigma_{23} - a_{21}\sigma_{13}$$

$$\sigma_{u'w'} = \sigma_{24} - a_{21}\sigma_{14}$$

$$\sigma_{v'w'} = \sigma_{34} - a_{31}\sigma_{14}$$

Thus, if we combine sample estimates of the variances and covariances of our observed variables with the least-squares estimates of the a's, we will generate estimates of the reduced-form disturbance variances and covariances. Putting these through the solution routine outlined earlier will yield estimates of structural coefficients and structural-form disturbance variances precisely the same as the estimates obtained by direct least-squares estimation of the structural equations themselves.

Comparing the path diagrams of the structural and reduced forms of the model may put some of these results in perspective:

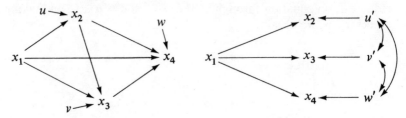

The diagrams are equivalent in one sense for, as we have shown, given the parameters (coefficients, variances, and covariances of disturbances) of one form, we may solve for the parameters of the other. Each diagram depicts a model with ten parameters. In the structural form we have:

one variance of the exogenous variable
six structural coefficients
three variances of disturbances

In the reduced form we have:

one variance of the exogenous variable
three reduced-form coefficients
three variances of reduced-form disturbances
three covariances among reduced form disturbances

The paths in the reduced-form diagram represent (typically) some combination of compound paths in the structural-form diagram. This fact is an instructive implication of our definitions of the a's (see Figure 4.1).

Thus, the reduced-form coefficients sum up the several direct and indirect paths through which the exogenous variable exerts its effects on each dependent variable. If one cared to know only the *total* effect of the exogenous variable on a dependent variable, the reduced-form coefficient tells the whole story. But if one is interested in how that effect comes about, the greater detail of the structural model is informative. After all, in the reduced form, a great deal of the "structure" is buried in the rather uninformative variances and covariances of the reduced-form disturbances.

Exercise. *Derive the reduced form for a two-equation model consisting only of the x_3-equation and the x_4-equation in the model just discussed. Both x_1 and x_2 are exogenous, so that the variances of both and their covariance as well must be assumed to arise from exogenous sources. Express the variances and covariances of the reduced-form disturbances in terms of structural coefficients and structural-form disturbance variances. Show how the direct and indirect effects of exogenous variables (insofar as these are explicit in the model) are summed up in reduced-form coefficients. Compare the number of parameters in the structural and reduced forms.*

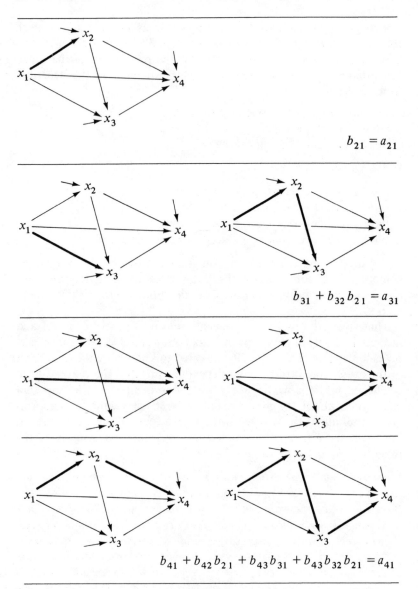

$$b_{21} = a_{21}$$

$$b_{31} + b_{32}b_{21} = a_{31}$$

$$b_{41} + b_{42}b_{21} + b_{43}b_{31} + b_{43}b_{32}b_{21} = a_{41}$$

FIG. 4.1. How structural coefficients combine into reduced form coefficients in a fully recursive model

We have managed to postpone to this point a matter that many sociologists consider—erroneously, we believe—to be the single most important feature of a model, the proportion of variation in each dependent variable that is "explained." Although this emphasis upon the misnamed "explanatory power" of a model is mistaken, there is a limited utility in the multiple-correlation statistic. We proceed to indicate how it fits into the account of recursive models offered here.

Suppose the structural coefficients of the model (page 52) have been estimated by OLS. This method minimizes the sum of squares of the sample residuals and thereby insures that the residual from an estimated regression equation is uncorrelated with the regressors in that equation. Hence we have $\sum x_1 \hat{u} = \sum x_2 \hat{v} = \sum x_1 \hat{v} = \sum x_3 \hat{w} = \sum x_2 \hat{w} = \sum x_1 \hat{w} = 0$, where the summation is over the entire sample and $\hat{u}$, $\hat{v}$, and $\hat{w}$ are the sample residuals that estimate the corresponding disturbances. This allows us to operate on the equations of the model, when the structural coefficients therein are replaced by their OLS estimates, in much the same way that we have hitherto worked with the model itself. Thus, *in the sample*, we have

$$x_2 = \hat{b}_{21} x_1 + \hat{u}$$
$$x_3 = \hat{b}_{32} x_2 + \hat{b}_{31} x_1 + \hat{v}$$
$$x_4 = \hat{b}_{43} x_3 + \hat{b}_{42} x_2 + \hat{b}_{41} x_1 + \hat{w}$$

(These are, in effect, the formulas for computing $\hat{u}$, $\hat{v}$, and $\hat{w}$, respectively.) Each variable will have been expressed as a deviation from its sample mean. Multiplying each equation through by the residual and the dependent variable, we find

$$\sum x_2 \hat{u} = m_{\hat{u}\hat{u}}$$
$$\sum x_3 \hat{v} = m_{\hat{v}\hat{v}}$$
$$\sum x_4 \hat{w} = m_{\hat{w}\hat{w}}$$
$$m_{22} = \hat{b}_{21} m_{12} + m_{\hat{u}\hat{u}}$$
$$m_{33} = \hat{b}_{32} m_{23} + \hat{b}_{31} m_{13} + m_{\hat{v}\hat{v}}$$
$$m_{44} = \hat{b}_{43} m_{34} + \hat{b}_{42} m_{24} + \hat{b}_{41} m_{14} + m_{\hat{w}\hat{w}}$$

The respective *coefficients of determination* for the three equations are

$$R^2_{2(1)} = 1 - \frac{m_{\hat{u}\hat{u}}}{m_{22}}$$

$$R^2_{3(21)} = 1 - \frac{m_{\hat{v}\hat{v}}}{m_{33}}$$

$$R^2_{4(321)} = 1 - \frac{m_{\hat{w}\hat{w}}}{m_{44}}$$

(The *multiple correlation* is the square root of R^2.) If we wished to put our results in the framework of standardized variables and path coefficients, we would proceed to note that

$$\hat{p}_{2u} = \sqrt{1 - R^2_{2(1)}}$$
$$\hat{p}_{3v} = \sqrt{1 - R^2_{3(21)}}$$
$$\hat{p}_{4w} = \sqrt{1 - R^2_{4(321)}}$$

which are the so-called residual paths.

It will be seen that the definition of R^2 rests on the distinction between variation "explained" by an equation in the model and the "unexplained" variation of that equation's dependent variable. It is sometimes suggested that the formula defining R^2 be used to effect a partitioning of the explained variation into portions due uniquely to the several determining causes. Thus, according to this suggestion, $R^2_{3(21)}$ (for example) would be allocated between x_1 and x_2 in proportion to $\hat{b}_{31} m_{13}$ and $\hat{b}_{32} m_{23}$, respectively (or, in the framework of path coefficients, $\hat{p}_{31} r_{13}$ and $\hat{p}_{32} r_{23}$). But the suggestion is mistaken. It is true that $\hat{b}_{31}$ estimates the direct effect of x_1 on x_3, but m_{13} reflects a mixture of effects arising from diverse sources. If we multiply through our estimate of the x_3-equation by x_1, we find

$$m_{13} = \hat{b}_{32} m_{12} + \hat{b}_{31} m_{11}$$

Hence

$$\hat{b}_{31} m_{13} = \hat{b}_{32} \hat{b}_{31} m_{12} + \hat{b}^2_{31} m_{11}$$

and the first term on the right is certainly not an unalloyed indicator of the role of x_1 alone in producing variation in x_3.

Indeed the "problem" of partitioning R^2 bears no essential relationship to estimating or testing a model, and it really does not add anything to our understanding of how a model works. The simplest recommendation—one which saves both work and worry—is to eschew altogether the task of dividing up R^2 into unique causal components. In a strict sense, it just cannot be done, even though many sociologists, psychologists, and other quixotic persons cannot be persuaded to forego the attempt.

Indeed the whole issue of what to make of a multiple correlation is clarified by noting that R^2 does not estimate any parameter of the model as such, but rather a parameter that depends upon both the model and the population in which the model applies. We have no reason to expect R^2 to be invariant across populations. If either σ_{ww} or σ_{11} (and, therefore, σ_{44}) changes in going to a new population, while the structural coefficients remain fixed, R^2 will be changed. This is a matter of special concern in the event that the value of σ_{11} is essentially under the investigator's control, according to whether he (for example) puts greater or lesser variance into the distribution of his experimental stimulus x_1, or samples disproportionately from different parts of the "natural" range of x_1. As we have seen, such a modification of σ_{11} will change all variances and covariances, as well as R^2, even though there is no change in structural coefficients—and there is no reason to expect them to be affected by either of these aspects of study design.

The real utility of R^2 is that it tells us something about the precision of our estimates of coefficients, since the standard error of a coefficient is a function, among other things, of R^2.

It is a mistake—the kind of mistake easily made by the novice—to focus too much attention on the magnitude of R^2. *Other things being equal*, it is, of course, true that one prefers a model yielding a high R^2 to one yielding a lower value. But the *ceteris paribus* clause is terribly important. Merely increasing R^2 by lengthening the list of regressors is no great achievement unless the role of those variables in an extended causal model is properly understood and correctly represented.

Suppose, for example, that in using the four-variable recursive model we have been studying, the investigator became dissatisfied with the low value of $R^2_{3(21)}$. It would be quite easy to get a higher value of R^2, for example, by running the regression of x_3 on x_4, x_2, and x_1, and reporting the value of $R^2_{3(421)}$. But this regression does not correspond to the causal ordering of the variables, which was required to be specified at the outset. It is, therefore, an exceedingly misleading statistic. (One does not often see the mistake in quite this crude a form. But naïvely regressing causes on effects is far from being unknown in the literature.)

Another way to raise $R^2_{3(21)}$ would be to introduce another variable, say x'_3, that is essentially an alternative measure of x_3, though giving slightly different results. The regression of x_3 on x'_3, x_2, and x_1 is then guaranteed to yield a high value of R^2.

Indeed, the best-known examples of very high correlations are those selected to convey the notion of "spurious correlation," "nonsense correlation in time series," or other kinds of artifact. This shows us that high values of R^2, in themselves, are not sufficient to evaluate a model as successful.

Before worrying too much about his R^2, therefore, the investigator does well to reconsider the entire specification of the model. If that specification cannot be faulted on other grounds, the R^2 as such is not sufficient reason to call it into question.

Exercise. *To conclude, for the time being, your study of fully recursive models, review the material on estimation and testing in Chapter 3 and restate the essential points so that they apply to the recursive model as expressed without standardization of variables (page 52).*

FURTHER READING

An example of a recursive sociological model presented in terms of both standardized and nonstandardized coefficients appears in Duncan (1969). Note that the more interesting conclusions were developed on the basis of the latter. On the questionable value of commonly used measures of "relative importance" or "unique contribution" of the several variables in an equation, see Ward (1969), Cain and Watts (1970), and Duncan (1970).

5

A Just-Identified
Nonrecursive Model

The model considered throughout this chapter is

$$x_3 = b_{31} x_1 + b_{34} x_4 + u$$

$$x_4 = b_{42} x_2 + b_{43} x_3 + v$$

For convenience, $E(x_j) = 0$, $j = 1, \ldots, 4$, and $E(u) = E(v) = 0$. However, we do *not* put the variables in standard form. Variables x_1 and x_2 are *exogenous*; their variances and their covariance are not explained within the model. Variables x_3 and x_4 are *jointly dependent* or *endogenous*; the purpose of the model is to explain the behavior of these variables. Variables u and v are, respectively, the *disturbances* in the x_3-equation and the x_4-equation. Their presence accounts for the fact that x_3 and x_4 are not fully explained by their explicit determining factors. The model will be operational only if we can assume that disturbances are uncorrelated with exogenous variables; hence the specification $E(x_1 u) = E(x_1 v) = E(x_2 u) = E(x_2 v) = 0$. *This is a serious assumption.* The research worker must carefully consider what circumstances would violate it and whether his theoretical understanding of the situation under study permits him to rule out such violations.

In contrast to the case of a fully recursive model, in the nonrecursive model the specification of zero covariances between disturbances and exogenous variables does not lead to either zero covariance between the two disturbances or zero covariance between the disturbance and each explanatory variable in an equation. To see this, let us multiply through each equation of the model by every variable in the model, and take expectations. To express the result of this operation in a convenient form, we will adopt the notation $E(x_j^2) = \sigma_{jj}$ where *sigma* with the repeated subscript refers to the (population) variance of x_j, and $E(x_h x_j) = \sigma_{hj}$ where (if $h \neq j$) *sigma* with two different subscripts refers to the (population) covariance of x_h and x_j. From the x_3-equation in the model we obtain, after multiplying through by x_1 and x_2:

$$E(x_1 x_3) = b_{31} E(x_1^2) + b_{34} E(x_1 x_4) + E(x_1 u)$$
$$E(x_2 x_3) = b_{31} E(x_1 x_2) + b_{34} E(x_2 x_4) + E(x_2 u)$$

or, since $E(x_1 u) = E(x_2 u) = 0$,

$$\sigma_{13} = b_{31} \sigma_{11} + b_{34} \sigma_{14}$$
$$\sigma_{23} = b_{31} \sigma_{12} + b_{34} \sigma_{24} \qquad \text{Set } (i)$$

However, in multiplying through the x_3-equation by endogenous variables and disturbances, not all covariances involving the disturbance drop out; we find:

$$\sigma_{33} = b_{31} \sigma_{13} + b_{34} \sigma_{34} + \sigma_{3u}$$
$$\sigma_{34} = b_{31} \sigma_{14} + b_{34} \sigma_{44} + \sigma_{4u}$$
$$\sigma_{3u} = b_{34} \sigma_{4u} + \sigma_{uu}$$
$$\sigma_{3v} = b_{34} \sigma_{4v} + \sigma_{uv} \qquad \text{Set } (ii)$$

Similarly, in multiplying through the x_4-equation by exogenous variables, we obtain:

$$\sigma_{14} = b_{42} \sigma_{12} + b_{43} \sigma_{13}$$
$$\sigma_{24} = b_{42} \sigma_{22} + b_{43} \sigma_{23} \qquad \text{Set } (iii)$$

But, in multiplying it through by endogenous variables and disturbances, we find:

$$\sigma_{34} = b_{42}\sigma_{23} + b_{43}\sigma_{33} + \sigma_{3v}$$

$$\sigma_{44} = b_{42}\sigma_{24} + b_{43}\sigma_{34} + \sigma_{4v}$$

$$\sigma_{4u} = b_{43}\sigma_{3u} + \sigma_{uv}$$

$$\sigma_{4v} = b_{43}\sigma_{3v} + \sigma_{vv} \qquad \text{Set } (iv)$$

Equations in Sets (i), (ii), (iii), and (iv) are population moment equations, analogous to but different in significant ways from those pertaining to recursive models. Note the lack of symmetry in the pattern of σ's on the right-hand side in Sets (i) and (iii), in contrast with the symmetric pattern of the normal equations on page 53. Note also—or carry out an *Exercise* to show this—that we cannot replace σ_{3u} by σ_{uu} or σ_{4v} by σ_{vv}, so that the simplification mentioned on page 53 for the recursive model is not available here. The population moment equations serve to express the relationships holding among the structural coefficients of the model (the b's) and the variances and covariances in the population under study. A number of important properties of the model are disclosed in studying these sets of equations.

Note, first, that if the b's in Set (i) are regarded as unknown and the σ's as known, it is possible to solve these two equations for the b's:

$$b_{31} = \frac{\sigma_{13}\sigma_{24} - \sigma_{14}\sigma_{23}}{\sigma_{11}\sigma_{24} - \sigma_{12}\sigma_{14}}$$

$$b_{34} = \frac{\sigma_{11}\sigma_{23} - \sigma_{12}\sigma_{13}}{\sigma_{11}\sigma_{24} - \sigma_{12}\sigma_{14}} \qquad \text{Set } (v)$$

In practice, of course, we would not know the population variances and covariances. However, if we replace the σ's by the corresponding sample moments, $m_{jj} = \sum (x_j^2)$ and $m_{hj} = \sum (x_h x_j)$, where the summation is over all sample observations on variables x_h and x_j, we obtain

$$\hat{b}_{31} = \frac{m_{13}m_{24} - m_{14}m_{23}}{m_{11}m_{24} - m_{12}m_{14}}$$

$$\hat{b}_{34} = \frac{m_{11}m_{23} - m_{12}m_{13}}{m_{11}m_{24} - m_{12}m_{14}} \qquad \text{Set } (vi)$$

This method of obtaining the estimates, $\hat{b}_{31}$ and $\hat{b}_{34}$, of the structural coefficients is termed instrumental variables. Here it is equivalent to the method of indirect least squares, for reasons that will become clear later. The method works here because the number of equations in Set (i), obtained by multiplying through the x_3-equation of the model by all exogenous variables, is just the same as the number of unknown structural coefficients. This fact is implied when we describe the model as "just identified" or "exactly identified" with respect to the x_3-equation. If Set (i) included more equations than structural coefficients, we would describe the x_3-equation as "overidentified"; if there were fewer moment equations in Set (i) than structural coefficients, the x_3-equation would be "underidentified" or "unidentified." In the case of overidentification, we replace instrumental variables (IV) or indirect least squares (ILS) by special methods of estimation. In the case of underidentification, estimation of structural coefficients is not possible.

Turning to the moment equations in Set (iii), we find that the x_4-equation of the model is likewise exactly identified. The solution for the b's is

$$b_{42} = \frac{\sigma_{14}\sigma_{23} - \sigma_{13}\sigma_{24}}{\sigma_{12}\sigma_{23} - \sigma_{13}\sigma_{22}}$$

$$b_{43} = \frac{\sigma_{12}\sigma_{24} - \sigma_{14}\sigma_{22}}{\sigma_{12}\sigma_{23} - \sigma_{13}\sigma_{22}} \qquad \text{Set } (vii)$$

IV estimates, $\hat{b}_{42}$ and $\hat{b}_{43}$, may be obtained as before by replacing the σ's with sample moments, m_{jj} and m_{hj}:

$$\hat{b}_{42} = \frac{m_{14}m_{23} - m_{13}m_{24}}{m_{12}m_{23} - m_{13}m_{22}}$$

$$\hat{b}_{43} = \frac{m_{12}m_{24} - m_{14}m_{22}}{m_{12}m_{23} - m_{13}m_{22}} \qquad \text{Set } (viii)$$

Further study of this model is facilitated by solving for its *reduced form*, in which each endogenous variable is represented as a function of exogenous variables and disturbances only. Each equation is substituted into the other one. That is, for x_4 in the x_3-equation we substitute the right-hand side of the x_4-equation; and for x_3 in the x_4-equation we substitute the right-hand side of the x_3-equation. After collecting terms, this algebra yields the two reduced-form equations:

$$x_3 = \frac{1}{1 - b_{34}b_{43}}(b_{31}x_1 + b_{34}b_{42}x_2 + u + b_{34}v)$$

$$x_4 = \frac{1}{1 - b_{34}b_{43}}(b_{43}b_{31}x_1 + b_{42}x_2 + b_{43}u + v)$$

Let us adopt the notation,

$$a_{31} = \frac{b_{31}}{(1 - b_{34}b_{43})}$$

$$a_{32} = \frac{b_{34}b_{42}}{(1 - b_{34}b_{43})}$$

$$a_{41} = \frac{b_{43}b_{31}}{(1 - b_{34}b_{43})}$$

$$a_{42} = \frac{b_{42}}{(1 - b_{34}b_{43})}$$

If we now multiply through each reduced-form equation by the exogenous variables, we obtain

$$\sigma_{13} = a_{31}\sigma_{11} + a_{32}\sigma_{12}$$

$$\sigma_{23} = a_{31}\sigma_{12} + a_{32}\sigma_{22} \qquad \text{Set } (ix)$$

and

$$\sigma_{14} = a_{41}\sigma_{11} + a_{42}\sigma_{12}$$

$$\sigma_{24} = a_{41}\sigma_{12} + a_{42}\sigma_{22} \qquad \text{Set } (x)$$

since terms like $b_{43}E(x_1 u)$ and $E(x_2 v)$ drop out.

It appears that the reduced-form parameters (the a's) are exact nonlinear functions of the structural coefficients (the b's), and vice versa. Indeed, the four expressions defining the a's could just as well be regarded as four equations in the unknown b's, so that if the a's were known we could solve for the b's by the following routine:

$$b_{34} = \frac{a_{32}}{a_{42}}$$

$$b_{43} = \frac{a_{41}}{a_{31}}$$

Set (xi)

whence

$$1 - b_{34}b_{43} = 1 - \frac{a_{32}a_{41}}{a_{31}a_{42}}$$

and

$$b_{31} = a_{31}\left(1 - \frac{a_{32}a_{41}}{a_{31}a_{42}}\right)$$

$$b_{42} = a_{42}\left(1 - \frac{a_{32}a_{41}}{a_{31}a_{42}}\right)$$

Set (xii)

Of course, the a's are not known, since they are functions of population variances and covariances, as shown by Sets (ix) and (x). But we could obtain estimates of the reduced-form parameters by replacing the σ's in Sets (ix) and (x) by corresponding sample moments; thus:

$$\hat{a}_{31} = \frac{m_{13}m_{22} - m_{12}m_{23}}{m_{11}m_{22} - m_{12}^2}$$

$$\hat{a}_{32} = \frac{m_{11}m_{23} - m_{12}m_{13}}{m_{11}m_{22} - m_{12}^2}$$

and

$$\hat{a}_{41} = \frac{m_{14}m_{22} - m_{12}m_{24}}{m_{11}m_{22} - m_{12}^2}$$

$$\hat{a}_{42} = \frac{m_{11}m_{24} - m_{12}m_{14}}{m_{11}m_{22} - m_{12}^2}$$

We see that these are precisely the same as the estimates we would obtain for the coefficients of the ordinary least squares (OLS) regressions of (respectively) x_3 on x_2 and x_1, and x_4 on x_2 and x_1, that is, of each endogenous variable on all exogenous variables.

Suppose we now replace the a's in Sets (xi) and (xii) by the corresponding OLS estimates, the $\hat{a}$'s. We will obtain estimates of the b's that are the very IV estimates presented earlier as Sets (vi) and $(viii)$; that is,

$$\hat{b}_{34} = \frac{\hat{a}_{32}}{\hat{a}_{42}}$$

$$\hat{b}_{43} = \frac{\hat{a}_{41}}{\hat{a}_{31}}$$

$$\hat{b}_{31} = \hat{a}_{31}\left(1 - \frac{\hat{a}_{32}\hat{a}_{41}}{\hat{a}_{31}\hat{a}_{42}}\right)$$

$$\hat{b}_{42} = \hat{a}_{42}\left(1 - \frac{\hat{a}_{32}\hat{a}_{41}}{\hat{a}_{31}\hat{a}_{42}}\right)$$

are the IV estimates of the b's. This fact may not be immediately obvious, but it is easily proved by algebraic substitutions.

Exercise.   *Carry out this algebra.*

The fact that the $\hat{b}$'s are obtained from the $\hat{a}$'s, which in turn are OLS estimates of reduced-form regression coefficients, justifies the name indirect least squares.

In addition to its possible use for purposes of estimation, the reduced form of the model is instructive in the way it displays the mechanisms through which the exogenous variables influence the endogenous variables. In this connection, study of the path diagram of the model is also instructive.

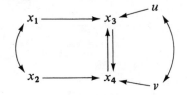

The covariance of x_1 and x_3 is obtained by multiplying through the reduced-form x_3-equation by x_1:

$$\sigma_{13} = \frac{b_{31}}{1 - b_{34}b_{43}}\sigma_{11} + \frac{b_{34}b_{42}}{1 - b_{34}b_{43}}\sigma_{12}$$

Note that there is a direct effect of x_1 on x_3:

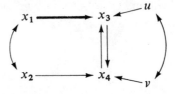

But another part of σ_{13} arises from the correlation (covariance) of x_1 with another cause (namely, x_2) of x_3, even though the latter works only indirectly:

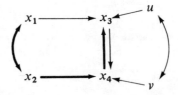

Note that we are reading the path diagram according to Sewall Wright's general principle, but are not using it as an algorithm for computing covariances. In particular, when actually calculating σ_{13} we must inflate both the direct path and the component due to a correlated cause by the factor $1/(1 - b_{34}b_{43})$. This is the "multiplier effect" in the model due to the "simultaneity" or "reciprocal causation" of the two endogenous variables. The covariance of x_2 and x_3 is

$$\sigma_{23} = \frac{b_{31}}{1 - b_{34}b_{43}} \sigma_{12} + \frac{b_{34}b_{42}}{1 - b_{34}b_{43}} \sigma_{22}$$

We see that it is produced by an indirect effect of x_2 on x_3:

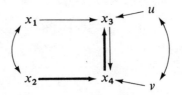

and by a component due to the correlation of x_2 with x_1:

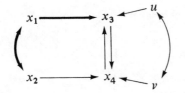

Again the multiplier effect comes into play for both components.

The same kinds of components can be found for the covariances of the two exogenous variables with x_4, shown below as they are obtained from the reduced-form x_4-equation:

$$\sigma_{24} = \frac{b_{42}}{1 - b_{34}b_{43}}\sigma_{22} + \frac{b_{43}b_{31}}{1 - b_{34}b_{43}}\sigma_{12}$$

$$\sigma_{14} = \frac{b_{42}}{1 - b_{34}b_{43}}\sigma_{12} + \frac{b_{43}b_{31}}{1 - b_{34}b_{43}}\sigma_{11}$$

Thus, σ_{24} arises from the direct effect:

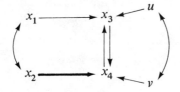

and the contribution due to a correlated cause, operating indirectly:

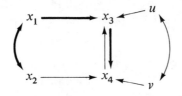

while σ_{14} arises from an indirect effect

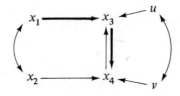

and the contribution of a correlated cause:

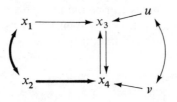

Again, all these components are seen to include the multiplier effect of the factor $1/(1 - b_{34}b_{43})$.

It is important to understand precisely why special methods, instead of the conventional statistical procedure of OLS, are required for estimating structural coefficients in nonrecursive models. In the x_3-equation (to take it as a typical example), the two explanatory variables on the right-hand side are x_1 and x_4. If one were to estimate the coefficients of the x_3-equation by OLS, therefore, he would obtain as the estimate of b_{31}

$$\frac{m_{13}m_{44} - m_{14}m_{34}}{m_{11}m_{44} - m_{14}^2}$$

and as the estimate of b_{34}

$$\frac{m_{11}m_{34} - m_{13}m_{14}}{m_{11}m_{44} - m_{14}^2}$$

As we have seen, in Sets (i) and (ii), the covariances of x_3 with these explanatory variables are

$$\sigma_{13} = b_{31}\sigma_{11} + b_{34}\sigma_{14}$$

$$\sigma_{34} = b_{31}\sigma_{14} + b_{34}\sigma_{44} + \sigma_{4u}$$

We may solve for the b's in terms of the σ's:

$$b_{31} = \frac{\sigma_{13}\sigma_{44} - \sigma_{14}\sigma_{34} + \sigma_{14}\sigma_{4u}}{\sigma_{11}\sigma_{44} - \sigma_{14}^2}$$

$$b_{34} = \frac{\sigma_{11}\sigma_{34} - \sigma_{13}\sigma_{14} - \sigma_{11}\sigma_{4u}}{\sigma_{11}\sigma_{44} - \sigma_{14}^2}$$

The implication is clear if we note that, apart from $\sigma_{14}\sigma_{4u}$ in the numerator, the expression for b_{31} is the population counterpart of the OLS estimator of b_{31}; and apart from $-\sigma_{11}\sigma_{4u}$ in the numerator, the expression for b_{34} is the population counterpart of the OLS estimator of b_{34}. Thus, even if our sample were infinitely large, so that we could form OLS estimators from population variances and covariances (instead of sample moments), the OLS estimates would be biased. Indeed, the OLS procedure would not estimate b_{31} but rather

$$b_{31} - \frac{\sigma_{14}\sigma_{4u}}{\sigma_{11}\sigma_{44} - \sigma_{14}^2}$$

and it would not estimate b_{34} but rather

$$b_{34} + \frac{\sigma_{11}\sigma_{4u}}{\sigma_{11}\sigma_{44} - \sigma_{14}^2}$$

Similarly we can show that OLS applied to the x_4-equation would estimate not b_{42} but rather

$$b_{42} - \frac{\sigma_{23}\sigma_{3v}}{\sigma_{22}\sigma_{33} - \sigma_{23}^2}$$

and, instead of b_{43}, it would estimate

$$b_{43} + \frac{\sigma_{22}\sigma_{3v}}{\sigma_{22}\sigma_{33} - \sigma_{23}^2}$$

The basic reason for the failure of OLS, then, is that not all the explanatory variables in the equation are uncorrelated with the disturbance. And this is inescapably so given the jointly dependent (simultaneous, reciprocally influencing) relationships of the endogenous variables of a nonrecursive model. For if $x_3 \rightarrow x_4$, then $x_u \rightarrow x_3$ implies that $x_u \rightarrow x_3 \rightarrow x_4$ will contribute a nonzero component to σ_{4u}. Similarly,

$x_v \to x_4 \to x_3$ will contribute a nonzero component to σ_{3v}. It could happen that one or the other of these is cancelled out by σ_{uv}, for in Sets (ii) and (iv) we find

$$\sigma_{3v} = b_{34}\sigma_{4v} + \sigma_{uv}.$$

and

$$\sigma_{4u} = b_{43}\sigma_{3u} + \sigma_{uv}.$$

Hence, if $\sigma_{uv} = -b_{34}\sigma_{4v}$, then $\sigma_{3v} = 0$; or if $\sigma_{uv} = -b_{43}\sigma_{3u}$, then $\sigma_{4u} = 0$. But for either of these to hold would be merely a coincidence, and for both to hold would be a rare coincidence indeed.

Exercise.   *Working from Sets (i), (ii), (iii), and (iv) in the spirit of Chapter 4, pages 53–55, show that the variances and covariances of the observable variables in the model studied in this chapter may be expressed as functions of (1) the variances and covariances of the exogenous variables, (2) a (nonlinear) combination of structural coefficients, and (3) variances and covariances of the disturbances. Verify Table 5.1.*

Table 5.1   Sources of Observable Variances and Covariances

Variance or covariance	Is a function of									
	σ_{11}	σ_{12}	σ_{22}	b_{31}	b_{42}	b_{34}	b_{43}	σ_{uu}	σ_{uv}	σ_{vv}
* σ_{11}	X	...	...	...	...	...	...	...	...	...
σ_{12}	...	X	...	...	...	...	...	...	...	...
σ_{22}	...	...	X	...	...	...	...	...	...	...
σ_{13}	X	X	...	X	X	X	X	...	...	...
σ_{14}	X	X	...	X	X	X	X	...	...	...
σ_{23}	...	X	X	X	X	X	X	...	...	...
σ_{24}	...	X	X	X	X	X	X	...	...	...
σ_{33}	X	X	X	X	X	X	X	X	X	X
σ_{34}	X	X	X	X	X	X	X	X	X	X
σ_{44}	X	X	X	X	X	X	X	X	X	X

* Exogenous

Discuss implications of the possibility that some, if not all, of the parameters across the top of the table are invariant across populations.

FURTHER READING

Chapters 5, 6, and 7 are essentially the elementary portions of the standard econometric presentation of simultaneous-equation models. Several advanced econometrics texts are listed among the references at the end of this book. More accessible presentations are available in Wonnacott and Wonnacott (1970, Part I), and Wallis (1973). The latter does, however, make use of matrices.

6

Underidentification and the Problem of Identification

Let us make one change in the model considered in Chapter 5 so that x_4 now depends on both exogenous variables. The new model is

$$x_3 = b_{31}x_1 + b_{34}x_4 + u$$
$$x_4 = b_{41}x_1 + b_{42}x_2 + b_{43}x_3 + v$$

The path diagram is

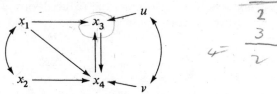

Multiplying through the x_3-equation by exogenous variables we obtain (as before)

$$\sigma_{13} = b_{31}\sigma_{11} + b_{34}\sigma_{14}$$
$$\sigma_{23} = b_{31}\sigma_{12} + b_{34}\sigma_{24}$$

Hence, we may estimate the structural coefficients of the x_3-equation by IV. (x_1 serves as its own instrument, whereas x_2 is the instrument for x_4, which cannot perform this role for itself because of its correlation with u.) The estimates are given in Set (vi) of Chapter 5.

Turning to the x_4-equation we note that only x_1 and x_2 are available as instrumental variables, since x_3 is correlated with v. Multiplying through by the instrumental variables we obtain:

$$\sigma_{14} = b_{41}\sigma_{11} + b_{42}\sigma_{12} + b_{43}\sigma_{13}$$

$$\sigma_{24} = b_{41}\sigma_{12} + b_{42}\sigma_{22} + b_{43}\sigma_{23}$$

We see that even if the σ's were known we could not solve uniquely for the b's, since there are three unknowns in only two equations. The x_4-equation of this model is $underidentified$. Note that the problem of identification is quite distinct from problems due to errors of sampling. We would be unable to estimate the structural coefficients in an underidentified equation even if we knew the population variances and covariances.

Another perspective on the identification problem is gained in examining the reduced form of the model. Substituting each equation into the other we obtain

$$x_3 = \frac{(b_{31} + b_{34}b_{41})x_1 + b_{34}b_{42}x_2 + u + b_{34}v}{1 - b_{34}b_{43}}$$

$$x_4 = \frac{(b_{41} + b_{43}b_{31})x_1 + b_{42}x_2 + b_{43}u + v}{1 - b_{34}b_{43}}$$

We may adopt new symbols for the reduced-form coefficients; their definitions serve to express the reduced-form coefficients in terms of the structural coefficients:

$$a_{31} = \frac{b_{31} + b_{34}b_{4i}}{1 - b_{34}b_{43}}$$

$$a_{32} = \frac{b_{34}b_{42}}{1 - b_{34}b_{43}}$$

$$a_{41} = \frac{b_{41} + b_{43}b_{31}}{1 - b_{34}b_{43}}$$

$$a_{42} = \frac{b_{42}}{1 - b_{34}b_{43}}$$

The analysis of the fully identified model (see page 72) carries over to the extent that we can estimate the a's by OLS. However, even if we knew the a's, we would not be able to solve for all the b's, since there are five unknowns in the four equations defining the a's. It does turn out that we can solve the reduced-form equations for $b_{34} = a_{32}/a_{42}$ and for $b_{31} = a_{31} - b_{34}a_{41}$. This corresponds to the fact that the x_3-equation is just identified, even though the x_4-equation is underidentified.

Thus one diagnosis of underidentification arises from study of the model's reduced form: If there are not enough reduced-form coefficients to define solutions for the structural coefficients, at least one of the equations of the model is underidentified.

A general counting rule is perhaps easier to apply. For each equation of a model count the number (G) of explanatory variables (variables on which the dependent variable depends directly, or which have causal arrows pointing directly to it). Then count the number (H) of variables available as instrumental variables; these will include all exogenous variables in the model and any other variables that are predetermined with respect to the particular equation. (In the simple nonrecursive models considered thus far, the only predetermined variables are, in fact, the strictly exogenous ones.) A necessary condition for identification is that $H \geq G$. (This is the so-called "order condition" for identification; but we shall not explain that term here.) If $H < G$, the equation is underidentified. For the x_4-equation in our illustrative model we find $G = 3$ (counting x_3, x_2, and x_1 as explanatory variables) and $H = 2$ (counting the exogenous variables x_1 and x_2 as instrumental variables); $H < G$, so that the x_4-equation is underidentified. The counting rule is necessary, but not strictly sufficient, although it usually suffices in practice, except for the kind of pathological model noted presently.

The sufficient condition for identification (the so-called "rank condition") is that each equation of a model be distinct from every other equation in the model and from all possible linear combinations of equations in the model. [We will not try to elucidate this statement, but simply refer the sufficiently highly motivated reader to the technical econometric literature, especially Christ (1966). However, we give below an example of how the condition may be violated.] If this condition is satisfied and if $H = G$, the equation is exactly or just identified; but if $H > G$, it is overidentified.

Note how this definition of overidentification applies to the recursive model studied in Chapter 3 (pages 44–50). The x_4-equation of that model included two explanatory variables, whereas three variables in the model were predetermined with respect to x_4. There was no need there to resort to instrumental variables not in the x_4-equation (indeed, as was indicated, it would be a mistake to do so), since each of the explanatory variables was, in fact, predetermined and could serve as its own instrument.

We consider later how to proceed in the case of overidentified non-recursive models, but note here only that both overidentified $(H > G)$ and just identified $(H = G)$ models are termed "identified."

Our present concern is how to recognize underidentification. We present a new example:

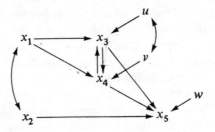

The equations of the model are

$$x_3 = b_{31}x_1 + b_{34}x_4 + u$$

$$x_4 = b_{41}x_1 + b_{43}x_3 + v$$

$$x_5 = b_{52}x_2 + b_{53}x_3 + b_{54}x_4 + w$$

We pause in the discussion of underidentification to observe that this model combines features of the two main kinds of models—recursive and nonrecursive. With regard to x_5 all the preceding variables are predetermined, and the specification on the disturbance of the x_5-equation is $E(x_j w) = 0, j = 1, \ldots, 4$.

Exercise. *Determine whether the x_5-equation is identified and, if it is, whether it is just identified or overidentified. If it is overidentified, determine the overidentifying restriction and suggest the appropriate methods* *of testing the restriction and of estimating the coefficients, assuming the overidentifying restriction is believed to obtain.*

With regard to x_3 and x_4, the model is nonrecursive, since these are jointly dependent variables. The specifications on their disturbances are $E(x_j u) = E(x_j v) = 0, j = 1, 2$, since both x_1 and x_2 are exogenous.

The model as a whole is *block-recursive*. The x_3- and x_4-equations comprise the first block; the x_5-equation by itself makes up the second block. The property of recursivity holds as between such blocks, the separability of which turns on the fact that they do not share any *endogenous* variables. (In the present example, x_3 and x_4 are endogenous with respect to the first two equations, but predetermined with respect to x_5.)

In our analysis of identification we focus on the x_3- and x_4-equations. Multiplying through by exogenous variables, we find

$$\left.\begin{aligned} \sigma_{13} &= b_{31}\sigma_{11} + b_{34}\sigma_{14} \\ \sigma_{23} &= b_{31}\sigma_{12} + b_{34}\sigma_{24} \end{aligned}\right\} \quad \text{from the } x_3\text{-equation}$$

$$\left.\begin{aligned} \sigma_{14} &= b_{41}\sigma_{11} + b_{43}\sigma_{13} \\ \sigma_{24} &= b_{41}\sigma_{12} + b_{43}\sigma_{23} \end{aligned}\right\} \quad \text{from the } x_4\text{-equation}$$

Taking the σ's as known and solving for the b's, we obtain

$$b_{31} = \frac{\sigma_{13}\sigma_{24} - \sigma_{14}\sigma_{23}}{\sigma_{11}\sigma_{24} - \sigma_{12}\sigma_{14}}$$

$$b_{34} = \frac{\sigma_{11}\sigma_{23} - \sigma_{12}\sigma_{13}}{\sigma_{11}\sigma_{24} - \sigma_{12}\sigma_{14}}$$

$$b_{41} = \frac{\sigma_{14}\sigma_{23} - \sigma_{13}\sigma_{24}}{\sigma_{11}\sigma_{23} - \sigma_{12}\sigma_{13}}$$

$$b_{43} = \frac{\sigma_{11}\sigma_{24} - \sigma_{12}\sigma_{14}}{\sigma_{11}\sigma_{23} - \sigma_{12}\sigma_{13}}$$

Now we observe a disconcerting feature of the solution: $b_{43} = 1/b_{34}$ whatever the values of the σ's, and similarly $b_{41} = -b_{31}/b_{34}$. So there is really only one set of coefficients that governs both of the equations. Or, more accurately, there really is only one equation, and whether we call it the x_3-equation or the x_4-equation is a matter of indifference.

Perhaps that should have been clear at the outset, for now we see that it is possible to rearrange the x_3-equation to read

$$x_4 = -\frac{b_{31}}{b_{34}}x_1 + \frac{1}{b_{34}}x_3 - \frac{1}{b_{34}}u$$

which is indistinguishable in form from the original x_4-equation. There is simply no way to tell whether we are estimating b_{41} or $-b_{31}/b_{34}$, whether we are estimating b_{43} or $1/b_{34}$.

Let us imagine a scenario—one with a basis in experience and not wholly fictitious. An investigator is working on a three-variable problem. He feels confident that x_1 precedes x_2 and x_3 in a causal ordering, but is uncertain which way the causal arrow runs between x_2 and x_3. That is, he is trying to choose between the models,

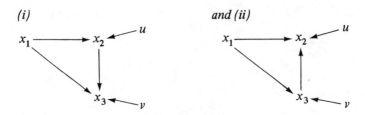

He resolves to let the question be decided by the data and specifies the nonrecursive model

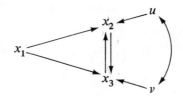

Thus, he reasons, if b_{23} is large and b_{32} is small, I will conclude that the predominance of the causation is in the direction $x_3 \rightarrow x_2$, so that model (ii) is preferred; if the opposite is true, I will decide for model (i). At this point, he sees that by the counting rule, both the x_2-equation and the x_3-equation are underidentified. Hence, he introduces an instrumental variable, x_0, and considers the following model, which by the counting rule appears to be identified:

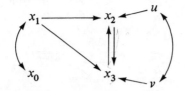

But we know already the end of this story. The hapless investigator works so hurriedly in making his computations (which go smoothly enough, offering no hint that anything is wrong) that he fails to notice the curious fact that $\hat{b}_{32} = 1/\hat{b}_{23}$, precisely. He does note, however, that $\hat{b}_{23}$ is large while $\hat{b}_{32}$ is small, and (without reporting this preliminary investigation to anyone in particular) in his further research treats the $x_2 \rightarrow x_3$ path as negligible.

Moral: Underidentification, not "causal inference," is achieved by "letting the data decide which way the causal arrow runs." The data cannot decide this matter, except in the context of a very strong theory, as is illustrated in the next exercise.

Exercise. *Imagine that the reasoning of the hypothetical investigator had been different. She began with a model in which both equations were underidentified:*

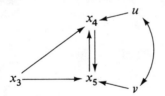

It then occurred to her to introduce an exogenous variable that appeared in the x_4-equation but not the x_5-equation:

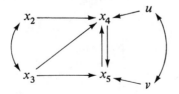

Show that this results in the x_5-equation being just identified, while the x_4-equation is still underidentified. If the investigator next introduced still another exogenous variable, which this time appears only in the x_5-equation, she would have the model

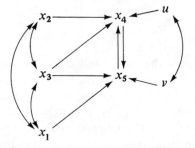

What is the status of each equation in this last model with respect to identification? What do you conclude about the kind of theory that is needed as a basis for specifying nonrecursive models in such a way that they are identified? What theories in sociology are known to you that persuasively provide such a basis?

Exercise. *Discuss the following model from the standpoint of identification:*

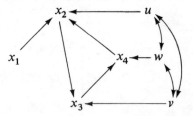

If one or more equations are underidentified, describe modifications of the model that would render all its equations identified.

The Aim of the Game

When the identification problem is presented in a purely formal way—as we have done here, for compactness—one's suspicions are certainly aroused that achieving identification is only a game. If your first sketch of a model turns out to be underidentified, just put in another variable in the "right" place and see if that shortcoming is remedied. But, of course, however simple "putting in another variable" may be in mathematical terms, it is a difficult undertaking in substantive terms. Our training in what passes for sociological theory tends to inculcate the healthy instinct to presume that "everything is connected to everything else." But a model in which this is true and in which all the connections are direct is an underidentified model—sometimes called, for rhetorical purposes, a "hopelessly underidentified" model.

Moreover, it is not enough just to "put in another variable," even if that variable is in the "right place." The additional variable(s), such as x_1 and x_2 in the exercise on page 87, must really belong in the model. Such variables must "make a difference" in the endogenous variable of the equation whose identifiability is in question, even though that difference is produced solely via indirect paths. From the standpoint of statistical estimation it must be the case that the variance in the endogenous variable produced (indirectly) by the exogenous variable(s) omitted from its equation is nontrivial. Looked at in this way (although the issue is too difficult to explore with our elementary methods), degree of identifiability may vary from weak to strong, whereas our formal analysis seemingly suggests that identification is an all-or-none proposition. For this reason, Klein (1962) advises us, "Identification cannot be cheaply achieved in any particular investigation by simply adding some weak or marginal variable to one of the relationships of a system. One must add something substantial and significant which had been previously neglected [p. 18]."

The identification problem with nonrecursive models is much the same as the problem of causal ordering with recursive models. You have to be able to argue convincingly that certain logically possible direct connections between variables are, in reality, nonexistent. Your theory must provide you with a secure basis for "sectoring" the world in such a way that the causal mechanisms of Equation 1 are really different from those operating in Equation 2 while still a different set of mechanisms comes into play in Equation 3, and so on. If the endogenous variables in all these equations are really just slightly different measures of the same thing—say, an individual's attitudes on three different but closely related issues—it is going to require a very subtle and elaborate theory indeed to produce distinct sets of determinants of those attitudes. If, by contrast, the first equation describes the behavior of labor, the second the behavior of management, and the third the behavior of government (or, respectively, the behaviors of the father, the mother, and the child), we may more easily argue that at least some of the causes involved in each equation do not appear in all the other equations.

Sociological studies involving serious efforts to construct nonrecursive models are still so few that no conclusion can be drawn as to the productivity of this approach. One can only offer conjectures, as already stated, concerning the kinds of problem that may prove amenable to study by such models. It does seem likely, however, that some modifications in our habits of theory construction—and not only in our practice of statistical analysis—will have to occur before many convincing examples of nonrecursive models are forthcoming. An investment in the study of the formal properties of such models amounts to making a wager as to the direction of development in the subject matter discipline in which one will work.

FURTHER READING

A comprehensive treatment of the identification problem is Fisher (1966); although it is heavily mathematical, a number of instructive points are formulated verbally. A discussion of the identifiability of a particular sociological model is found in Henry and Hummon (1971) with reply by Woelfel and Haller (1971). For an appreciation of Sewall Wright's long-neglected contribution to the identification problem see Goldberger (1972a). The classic paper employing modern nomenclature in expounding the identification problem was published by Koopmans in 1949; it is reprinted in Blalock (1971, Chap. 9). Criteria for "good" instrumental variables are suggested by Fisher (in Blalock, 1971, pages 260ff.).

7

Overidentification in a Nonrecursive Model

Let us enlarge the model considered in Chapter 5. We assume there are three exogenous variables, and their direct effects on the two jointly dependent variables are as shown in the path diagram:

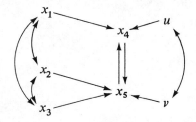

The model, therefore, is:

$$x_4 = b_{41}x_1 + b_{45}x_5 + u$$
$$x_5 = b_{52}x_2 + b_{53}x_3 + b_{54}x_4 + v$$

with the usual specification on the disturbances. Application of the counting rule (page 83) suggests that the x_5-equation is just identified (there are three explanatory variables in that equation and three exogenous variables in the model as a whole). The x_4-equation is overidentified (there are only two explanatory variables in this equation).

Multiplying through by exogenous variables, we obtain

$$\left.\begin{array}{l} \sigma_{14} = b_{41}\sigma_{11} + b_{45}\sigma_{15} \\[4pt] \sigma_{24} = b_{41}\sigma_{12} + b_{45}\sigma_{25} \\[4pt] \sigma_{34} = b_{41}\sigma_{13} + b_{45}\sigma_{35} \end{array}\right\} \quad \text{from the } x_4\text{-equation}$$

$$\left.\begin{array}{l} \sigma_{15} = b_{52}\sigma_{12} + b_{53}\sigma_{13} + b_{54}\sigma_{14} \\[4pt] \sigma_{25} = b_{52}\sigma_{22} + b_{53}\sigma_{23} + b_{54}\sigma_{24} \\[4pt] \sigma_{35} = b_{52}\sigma_{23} + b_{53}\sigma_{33} + b_{54}\sigma_{34} \end{array}\right\} \quad \text{from the } x_5\text{-equation}$$

We see that the IV method is available for estimating coefficients in the x_5-equation. The estimates are obtained by solving the following set of normal equations for the $\hat{b}$'s:

$$m_{15} = \hat{b}_{52}m_{12} + \hat{b}_{53}m_{13} + \hat{b}_{54}m_{14}$$

$$m_{25} = \hat{b}_{52}m_{22} + \hat{b}_{53}m_{23} + \hat{b}_{54}m_{24}$$

$$m_{35} = \hat{b}_{52}m_{23} + \hat{b}_{53}m_{33} + \hat{b}_{54}m_{34}$$

The situation is not so straightforward for the overidentified x_4-equation. The overidentifying restriction implies that

$$\overset{(i)}{} \qquad\qquad \overset{(ii)}{} \qquad\qquad \overset{(iii)}{}$$

$$b_{41} = \frac{\sigma_{14}\sigma_{25} - \sigma_{24}\sigma_{15}}{\sigma_{11}\sigma_{25} - \sigma_{12}\sigma_{15}} = \frac{\sigma_{14}\sigma_{35} - \sigma_{34}\sigma_{15}}{\sigma_{11}\sigma_{35} - \sigma_{13}\sigma_{15}} = \frac{\sigma_{24}\sigma_{35} - \sigma_{34}\sigma_{25}}{\sigma_{12}\sigma_{35} - \sigma_{13}\sigma_{25}}$$

and

$$b_{45} = \frac{\sigma_{11}\sigma_{24} - \sigma_{14}\sigma_{12}}{\sigma_{11}\sigma_{25} - \sigma_{12}\sigma_{15}} = \frac{\sigma_{11}\sigma_{34} - \sigma_{13}\sigma_{14}}{\sigma_{11}\sigma_{35} - \sigma_{13}\sigma_{15}} = \frac{\sigma_{12}\sigma_{34} - \sigma_{13}\sigma_{24}}{\sigma_{12}\sigma_{35} - \sigma_{13}\sigma_{25}}$$

(Although there are several equalities here, they are redundant. There is actually only one overidentifying restriction.) We might estimate these b's by replacing the σ's with sample moments in any one of these solutions. Note that neither solution (i), (ii), nor (iii) leads to an OLS estimate, in contrast to the result for the overidentified equation in a recursive model (page 46). (We already know, in any event, that OLS does not yield unbiased estimates in nonrecursive models, however large the sample may be.) If we replace the σ's by sample moments in the foregoing solutions, we will, in general, obtain three different pairs of values for the estimated b's. Because of sampling error the equalities among the solutions will be only approximate, not exact, even if the model—or, in particular, its overidentifying restriction—is true. The essence of the overidentified case, then, is that there are "too many" distinct estimates of the structural coefficients. It is not obvious how to choose the best one from among them, or how to reconcile them. It might seem plausible to average the estimates. In a sense, this is what is done by the method that will be described later on. But the appropriate average is not a simple, unweighted mean of the three estimates.

Since a direct application of the IV method does not work for an overidentified equation (there are "too many" instrumental variables and no firm basis for choosing among them), we look for help in another direction, by studying the reduced form of the model. We find

$$x_4 = a_{41}x_1 + a_{42}x_2 + a_{43}x_3 + u'$$

$$x_5 = a_{51}x_1 + a_{52}x_2 + a_{53}x_3 + v'$$

where the reduced-form coefficients and disturbances are the following functions of the structural-form coefficients and disturbances:

$$a_{41} = \frac{b_{41}}{1 - b_{45}b_{54}}$$

$$a_{42} = \frac{b_{45}b_{52}}{1 - b_{45}b_{54}}$$

$$a_{43} = \frac{b_{45}b_{53}}{1 - b_{45}b_{54}}$$

$$a_{51} = \frac{b_{54}b_{41}}{1 - b_{45}b_{54}}$$

$$a_{52} = \frac{b_{52}}{1 - b_{45}b_{54}}$$

$$a_{53} = \frac{b_{53}}{1 - b_{45}b_{54}}$$

$$u' = \frac{u + b_{45}v}{1 - b_{45}b_{54}}$$

$$v' = \frac{b_{54}u + v}{1 - b_{45}b_{54}}$$

Exercise.   *Using techniques similar to those in Chapters 4 and 5, verify these expressions. Find the variances of u′ and v′ and their covariance in terms of structural coefficients and variances and covariances of the structural-form disturbances. Remember that, in the nonrecursive case, $\sigma_{uv} = 0$ does not follow from the usual specification on the structural disturbances.*

It is apparent that our work in deriving the reduced-form coefficients has not solved our problem immediately. We find that $a_{42}/a_{52} = b_{45}$ but also $a_{43}/a_{53} = b_{45}$. If the model is true, both equalities must hold, so that

$$\frac{a_{42}}{a_{52}} = \frac{a_{43}}{a_{53}} \quad \text{or} \quad a_{42}a_{53} = a_{43}a_{52}$$

(This is another way of expressing the overidentifying restriction.) But, in practice, we do not know the a's and can only hope to secure estimates of them. Suppose we adopted as our estimates of the a's the OLS regression coefficients of x_4 on x_1, x_2, and x_3, and x_5 on x_1, x_2, and x_3, taking advantage of the fact (which the reader should verify) that covariances of u' and v' with the three exogenous variables are all zero. There is nothing about the OLS method which guarantees that estimates of coefficients in two different equations, estimated independently, will satisfy exactly the proportionality just cited. The best we could hope for is that

$$\frac{\hat{a}_{42}}{\hat{a}_{52}} \cong \frac{\hat{a}_{43}}{\hat{a}_{53}}$$

(where $\cong$ means "approximately equal to"). But this would leave us in the position of having to choose between the two distinct estimates

$$\hat{b}_{45}^{(1)} = \frac{\hat{a}_{42}}{\hat{a}_{52}}$$

and

$$\hat{b}_{45}^{(2)} = \frac{\hat{a}_{43}}{\hat{a}_{53}}$$

or otherwise reconciling the two. But there is no obvious way to do this, unless the reader, whose patience has by now worn thin, considers it "obvious" that it is good enough to "split the difference." We would still have to harass the distraught fellow, however, by insisting that there is more than one way to "split the difference"—for example, by reconciling different estimates obtained via the reduced form or by reconciling those obtained on the IV approach.

We can, however, put our OLS estimates of reduced form coefficients to good use for the purpose at hand. They will serve as "first stage regression" coefficients for the method known as two-stage least squares (2SLS).

We are working on the overidentified x_4-equation, and it is the presence of x_5 in that equation that occasions much of our difficulty. To finesse that source of difficulty, we proceed as follows. Define $\hat{x}_5$ as

$$\hat{x}_5 = \hat{a}_{51}x_1 + \hat{a}_{52}x_2 + \hat{a}_{53}x_3$$

where the $\hat{a}$'s are the OLS estimates of coefficients in the reduced-form x_5 equation. We have then

$$x_5 = \hat{x}_5 + \hat{v}'$$

where $\hat{v}'$ is an estimate, calculated as the sample residual from the estimated regression equation, of the reduced-form disturbance v'. We substitute this expression for x_5 into the x_4-equation of the model, obtaining

$$x_4 = b_{41}x_1 + b_{45}\hat{x}_5 + b_{45}\hat{v}' + u$$

Let us multiply through by the two explanatory variables and take expectations:

$$E(x_1 x_4) = b_{41}E(x_1^2) + b_{45}E(x_1\hat{x}_5) + b_{45}E(x_1\hat{v}') + E(x_1 u)$$

$$E(\hat{x}_5 x_4) = b_{41}E(\hat{x}_5 x_1) + b_{45}E(\hat{x}_5^2) + b_{45}E(\hat{x}_5\hat{v}') + E(\hat{x}_5 u)$$

We must look closely at the four terms involving disturbances. First, $E(x_1 u) = 0$ by the original specification of the model. Second, $E(x_1 \hat{v}')$ must be zero, since $\hat{v}'$ is the sample residual from a regression in which x_1 is one of the independent variables. It is a property of OLS that each regressor has a covariance of zero, identically, with the sample residual. Third, for much the same reason $E(\hat{x}_5 \hat{v}') = 0$; for $\hat{x}_5$ is the value of the dependent variable calculated from a regression equation, and $\hat{v}'$ is the residual from that same regression. The OLS method ensures that the covariance of the two is identically zero.

The situation is messier with respect to the last term, $E(\hat{x}_5 u)$. We recall that, by definition,

$$\hat{x}_5 = \hat{a}_{51} x_1 + \hat{a}_{52} x_2 + \hat{a}_{53} x_3$$

Suppose, for the moment, that we knew the actual values of the a's in the population and did not have to use the $\hat{a}$'s. Then we could compute a slightly different quantity,

$$x_5^* = a_{51} x_1 + a_{52} x_2 + a_{53} x_3$$

If we then considered $E(x_5^* u)$ we would find that its value is

$$a_{51} E(x_1 u) + a_{52} E(x_2 u) + a_{53} E(x_3 u) = 0$$

The a's can be written to the left of the expectation sign since they are constants. (The same is not true of the $\hat{a}$'s; they are random variables that vary from one sample to another.) And, of course, each of the expectations, $E(x_h u) = 0$, $h = 1$, 2, 3, since x_1, x_2, and x_3 are exogenous variables.

All this is very nice, but $\hat{x}_5$ is not the same as x_5^*; it is only an estimate of x_5^*. Here, we must appeal to some statistical theory that lies beyond the scope of this exposition. While we may only write $\hat{x}_5 \cong x_5^*$, the error in the approximation will diminish, on the average, as we take larger and larger samples. In the limit, as the sample gets indefinitely large, the probability that $\hat{x}_5$ differs from x_5^* by more than any prespecified amount tends to zero. Replacing x_5^* by $\hat{x}_5$ in the expectation $E(x_5^* u)$, therefore, we may write

$$E(\hat{x}_5 u) \cong 0$$

and the error in the approximation gets smaller, on the average, the larger the sample. Hence, $E(\hat{x}_5 u)$ is "asymptotically equal" to zero. The approximation involved in taking it to be identically zero is of the same kind that we use whenever we employ "large-sample statistics" or "asymptotic estimators."

We have presented a long argument to the effect that, when working with a "large" sample, we are justified in dropping the last two terms in the expressions for $E(x_1 x_4)$ and $E(\hat{x}_5 x_4)$ previously given. We find, therefore, that

$$E(x_1 x_4) = b_{41} E(x_1^2) + b_{45} E(x_1 \hat{x}_5)$$

$$E(\hat{x}_5 x_4) \cong b_{41} E(\hat{x}_5 x_1) + b_{45} E(\hat{x}_5^2)$$

We see that if the " $\cong$ " is replaced by " $=$," and the several expectations by the corresponding sample moments, we will produce OLS estimates, $\hat{b}_{41}$ and $\hat{b}_{45}$, by simply regressing x_4 on x_1 and $\hat{x}_5$, where $\hat{x}_5$ is calculated from the result of the first-stage regression and the two $\hat{b}$'s are estimated in this second-stage regression. What we do, in effect, is replace x_5 in the x_4-equation by the estimate of x_5 given by its regression on *all* the exogenous variables in the model, and then use OLS on this revised equation.

We might equally well estimate the x_5-equation by 2SLS. In that event, we would calculate the first-stage OLS regression of x_4 on x_1, x_2, and x_3; compute the calculated value ($\hat{x}_4$) of x_4 from that regression; replace x_4 in the x_5-equation by $\hat{x}_4$; and estimate the coefficients in the x_5-equation by OLS regression of x_5 on x_2, x_3, and $\hat{x}_4$. It turns out that the 2SLS estimates obtained for the just-identified x_5-equation are the same as the IV estimates. The fact that 2SLS and IV give the same result in the case of any equation that is just identified may be seen as a heuristic justification for the approximation used in deriving the 2SLS method.

Exercise. *What if the x_5-equation were underidentified; specifically, what if it were specified as*

$$x_5 = b_{51}x_1 + b_{52}x_2 + b_{53}x_3 + b_{54}x_4 + v$$

could we then estimate the coefficients by 2SLS, *where the second-stage regression is x_5 on x_1, x_2, x_3, and $\hat{x}_4$?*

Answer: No, because $\hat{x}_4$ is a weighted sum (with $\hat{a}$'s as weights) of x_1, x_2, and x_3, while all three of these variables appear elsewhere in the second-stage regression equation. Our efforts to calculate OLS estimates would fail because of "singularity." If this form of pathology is not known to you, ask your teacher to explain it.

Conclusion: Neither 2SLS nor any other method of estimation is a cure for underidentification, because the identification problem does not arise from sampling errors but from difficulties of a logical kind.

We have tried only to sketch the logic of 2SLS and not to describe efficient computational procedures. We do not, moreover, deal with tests of hypotheses about the individual structural coefficients. The standard errors required for such tests are produced by any good computer program for calculating the estimated 2SLS coefficients themselves. The intent of our discussion was to show the reader that some special method of estimation is both required and feasible whenever a model contains one or more overidentified equations.

There are several other statistically efficient methods besides 2SLS for estimating overidentified equations or, in the case of some methods, for estimating all equations in the model at once. All of these methods are more complex, conceptually and computationally, than 2SLS. Therefore, the well-motivated reader must be referred to the advanced textbooks of econometrics, of which there are several excellent ones (Johnston, Goldberger, Christ, Malinvaud, Theil, Kmenta, among others). For all the interest in these methods, most empirical work uses 2SLS, which is quite flexible in applications and which appears to have a certain robustness in the face of the practical difficulties that always arise in a serious piece of empirical work.

From estimation, we turn to some cursory remarks on the problem of testing overidentifying restriction(s). Formal procedures of statistical inference have been proposed in connection with this problem. But these procedures are little used in practice, and some questions remain about their purely statistical properties.

It is easy to see one reason why the outcome of any such test may not be highly instructive. First, suppose the model passes the test; one is not required to reject the null hypothesis which asserts the overidentifying restriction(s) to be true. But this outcome, of course, does not guarantee that the model is true; it only provides some reassurance to the investigator who thinks he has other, adequate reasons for believing it to be true.

Second, suppose the null hypothesis must be rejected. One then concludes, with only a small probability of being mistaken, that "something" about the overidentifying restriction(s) is wrong. In the example used throughout this discussion, the overidentifying restriction takes the simplest possible form; it asserts the equality of two ratios of reduced-form coefficients: $a_{42}/a_{52} = a_{43}/a_{53}$. In more highly overidentified models there will be several such conditions. But the rejection of the null hypothesis only tells one that "something" is wrong, not what in particular is likely to be wrong.

This is true even in our simple example. If we must reject the overidentifying restriction, how may we remedy the situation? If we draw in a causal arrow, $x_2 \to x_4$, and thereby revise the x_4-equation to read,

$$x_4 = b_{41}x_1 + b_{42}x_2 + b_{45}x_5 + u$$

this equation, as well as the x_5-equation, will be just identified. There will be no overidentifying restriction(s) and thus no way of rejecting the model for failure to fit the overidentifying restriction(s). But exactly the same thing will be true if, instead, we put in the arrow, $x_3 \to x_4$, so that the x_4-equation will read

$$x_4 = b_{41}x_1 + b_{43}x_3 + b_{45}x_5 + u$$

The result of the original test of the overidentifying restriction is of no help in deciding which (if not some other) route to take. A formal analysis can only reveal formal conditions that a good model must satisfy (or satisfy approximately). Whether it is really any good must be determined on substantive grounds, with the guidance of the best theory available.

Exercise.   *What happens if we revise the x_4-equation to include both x_2 and x_3?*

Exercise.   *Enumerate all possible two-equation nonrecursive models comprising two endogenous and three exogenous variables, each equation of every model being just identified. Let every model have the same set of endogenous variables (say, x_4 and x_5) and the same set of exogenous variables $(x_1, x_2, \text{ and } x_3)$. What possibility, if any, do you see for letting the choice of one from among this list of models be decided by the results of a statistical analysis?*

FURTHER READING

An overidentified sociological model is estimated by 2SLS in Duncan, Haller, and Portes (in Blalock, 1971, Chapter 13), but the presentation is unnecessarily complicated by the use of standardized variables.

8

Specification Error

The title of this chapter, like much of the nomenclature and content of this book, derives from the literature on econometrics. It is quite a useful euphemism for what in a blunter language would be called "using the wrong model." There are many more wrong models than right ones, so that specification error is very common, though often not recognized and usually not easily recognizable.

The main message of the entire book would be missed if the reader did not understand by now that we can seldom be sure we have the right model, although we can sometimes be nearly certain, on the basis of empirical evidence, that we have been using the wrong one. Indeed, it would require no elaborate sophistry to show that we will never have the "right" model in any absolute sense. Hence, we shall never be able to compare one of our many wrong models with a definitively right one. Is the matter of specification error beyond rational discussion, therefore, even if we have no difficulty in "taking about" it while failing to "discuss" it?

As the term will be used here, analysis of specification error relates to a rhetorical strategy in which we suggest a model as the "true" one for sake of argument, determine how our working model differs from it and what the consequences of the difference(s) are, and thereby get some sense of how important the mistakes we will inevitably make may be.

Sometimes it is possible to secure genuine comfort by this route. For example, in using an estimation method like 2SLS, where we estimate the coefficients of just one equation in the model at a time, it turns out that having specification errors in the other equations does not matter, if only we have designated correctly the predetermined variables that are excluded from the particular equation being estimated. (This is fairly "obvious" from a review of how this method goes, and no further proof is offered here.) It could happen (if one is lucky) that the particular equation is the crucial one for the light that its coefficients shed on the theory we are working with. Hence, this partial insulation from the effects of specification error can be welcome.

A similar theorem was stated for recursive models in Chapter 3 (page 43). In estimating path (or structural) coefficients in one equation, we do not harm the results by having an erroneous causal ordering of the variables in "prior" equations. Of course, we do harm to the results for the model as a whole, so it is well not to take too much comfort from this theorem.

It should not be supposed, however, that a mistake in regard to the causal ordering is the only form of specification error. To illustrate the contrary, suppose we are working with three variables and have a firm basis for the causal ordering, $x_1 \rightarrow x_2 \rightarrow x_3$. Can we be sure that it is safe to take

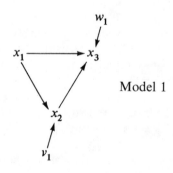

Model 1

as our model (especially since it subsumes the case in which any one of the structural coefficients is zero)?

Imagine the true state of affairs is

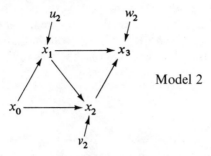

Model 2

(The disturbances carry subscripts because v and w in Model 2 are not the same as v and w in Model 1.) You should be able to see at a glance that Model 1 is faulty. Even though x_1 is causally prior to x_2 and x_3, it is not legitimate (in the light of "true" Model 2) to treat x_1 by itself as the sole exogenous variable. To demonstrate the specification error more rigorously and to discover its precise consequences requires some algebra of the kind we have been doing throughout our study.

Exercise. *Write down the equations for each model and from them derive expressions for the variances and covariances of all dependent variables. Set up the formulas for OLS estimates of structural coefficients.*

We find that the structural coefficients and OLS estimates thereof are the same for the x_3-equation in Model 2 as in Model 1. For the x_2-equation, however, we find that Model 1 (with the usual specification on the disturbance) implies $b_{21} = \sigma_{12}/\sigma_{11}$. But, from "true" Model 2 we see (given the results of the Exercise) that

$$\frac{\sigma_{12}}{\sigma_{11}} = \frac{b_{21}\sigma_{11} + b_{20}\sigma_{01}}{\sigma_{11}}$$

$$= b_{21} + b_{20}\frac{\sigma_{01}}{\sigma_{11}}$$

Hence, if we use the "false" estimator,

$$\hat{b}_{21}^{F} = \frac{m_{12}}{m_{11}}$$

we shall be estimating *not* b_{21} but a quantity that differs from b_{21} by an amount $b_{20}\sigma_{01}/\sigma_{11}$, which will be positive or negative according to whether b_{20} and σ_{01} have the same or different signs. Indeed, since $\sigma_{01} = b_{10}\sigma_{00}$ and $\sigma_{11} = b_{10}^2\sigma_{00} + \sigma_{u_2u_2}$ (you should use results of the exercise to verify this claim), we see that the bias in $\hat{b}_{21}^F$ can be written

$$b_{20}b_{10}\frac{\sigma_{00}}{b_{10}^2\sigma_{00} + \sigma_{u_2u_2}}$$

so that its sign depends on whether b_{20} and b_{10} have like or different signs (the remaining terms in the expression being intrinsically positive). We also see that if either b_{20} or b_{10} is zero, the bias is nil. (This agrees with the results of common sense in comparing the two path diagrams—or, if it does not, you need further to develop your common sense about these diagrams.) If the investigator can offer plausible considerations favoring the view that one or the other of these coefficients is "small," he may take comfort in the implication that the bias in his estimator $(\hat{b}_{21}^F)$ is small or, equivalently, that Model 1 is "nearly true." Such comfort would be prized in the realistic situation where Model 2 is "known" to be true, but measurements on x_0 are not available, whether by reason of some defect in the study design or because x_0 is not an observable variable.

Suppose, however, that x_0, though not observed, must be assumed to have substantial effects on both x_1 and x_2. What then is our recourse? We can only adjust to this unsatisfactory state of affairs by substituting for (false) Model 1 the following revised version:

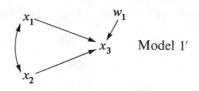

Model 1'

The fact that models like Model 1' are intrinsically less informative than those like Model 1 (if only the latter were true!) has been remarked in Chapter 3 (pages 36–43).

There is another instructive way to describe the specification error.

Let us compare the x_2-equations of the two models.

$$x_2 = b_{21}x_1 + v_1 \qquad \text{(Model 1)}$$

$$x_2 = b_{21}x_1 + b_{20}x_0 + v_2 \qquad \text{(Model 2)}$$

Now, we can render Model 1 "true" by resolving to think of b_{21} as the same coefficient, regardless of which model it appears in, and noting that this requires us to recognize a relationship between the disturbances in the two models, given by

$$v_1 = b_{20}x_0 + v_2$$

Evaluating $E(x_1 v_1)$ we find

$$E(x_1 v_1) = b_{20}E(x_0 x_1) + E(x_1 v_2)$$

$$= b_{20}\sigma_{01}$$

since in "true" Model 2 the usual specification on the disturbance would be $E(x_1 v_2) = 0$. But it is precisely this "usual" specification that we cannot legitimately adopt for Model 1, given the truth of Model 2. Another way to put the whole issue, then, is that in Model 1 (as we know from comparing it with Model 2) we must *not* assume $E(x_1 v_1) = 0$. But with that assumption not available, we cannot rely on the OLS estimate of b_{21} in the framework of Model 1. We come full circle back to the statement in Chapter 1 (page 5), "The assumption that an explanatory or causal variable is uncorrelated with the disturbance must always be weighed carefully." Perhaps the meaning of "to weigh carefully" will be a little clearer now than it was seven chapters ago.

Let us make use of this alternative way of describing specification error in attacking another illustrative problem. Suppose, now, the "true" model is

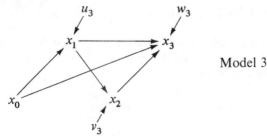

Model 3

Studying Model 3 after the fashion of Chapter 4 we find among other things:

$$\sigma_{01} = b_{10}\sigma_{00}$$

$$\sigma_{11} = b_{10}^2\sigma_{00} + \sigma_{u_3u_3}$$

$$\sigma_{02} = b_{21}b_{10}\sigma_{00}$$

$$\sigma_{12} = b_{21}b_{10}^2\sigma_{00} + b_{21}\sigma_{u_3u_3}$$

$$\sigma_{22} = b_{21}^2b_{10}^2\sigma_{00} + b_{21}^2\sigma_{u_3u_3} + \sigma_{v_3v_3}$$

Exercise. *Verify these results.*

Now, we propose to study Model 1 in light of true Model 3. But, this time, we make explicit the fact that $E(x_1 w_1) \neq 0$. That is, we take the x_3-equation from Model 1,

$$x_3 = b_{32}x_2 + b_{31}x_1 + w_1$$

and, after observing that if Model 3 is true,

$$w_1 = b_{30}x_0 + w_3$$

rewrite the x_3-equation as

$$x_3 = b_{32}x_2 + b_{31}x_1 + (b_{30}x_0 + w_3)$$

Now we multiply through by x_2 and x_1:

$$\sigma_{23} = b_{32}\sigma_{22} + b_{31}\sigma_{12} + (b_{30}\sigma_{02})$$

$$\sigma_{13} = b_{32}\sigma_{12} + b_{31}\sigma_{11} + (b_{30}\sigma_{01})$$

taking advantage of the specification in Model 3 that $E(x_1 w_3) = E(x_2 w_3) = 0$. Just to make the next step easier to follow, these two equations are rewritten

$$(\sigma_{23} - b_{30}\sigma_{02}) = b_{32}\sigma_{22} + b_{31}\sigma_{12}$$

$$(\sigma_{13} - b_{30}\sigma_{01}) = b_{32}\sigma_{12} + b_{31}\sigma_{11}$$

We solve for b_{32}:

$$b_{32} = \frac{\sigma_{11}(\sigma_{23} - b_{30}\sigma_{02}) - \sigma_{12}(\sigma_{13} - b_{30}\sigma_{01})}{\sigma_{11}\sigma_{22} - \sigma_{12}^2}$$

$$= \frac{\sigma_{11}\sigma_{23} - \sigma_{12}\sigma_{13}}{\sigma_{11}\sigma_{22} - \sigma_{12}^2} + b_{30}\frac{\sigma_{01}\sigma_{12} - \sigma_{02}\sigma_{11}}{\sigma_{11}\sigma_{22} - \sigma_{12}^2}$$

But the second term on the right vanishes, as will be seen upon evaluating the numerator, using the expressions for the σ's already exhibited. This shows that the OLS estimator of b_{32} from Model 1 is unbiased, for if x_3 is regressed on x_2 and x_1 the estimator is

$$\hat{b}_{32} = \frac{m_{11}m_{23} - m_{12}m_{13}}{m_{11}m_{22} - m_{12}^2}$$

On the other hand, when we solve for the other structural coefficient, we obtain

$$b_{31} = \frac{\sigma_{22}\sigma_{13} - \sigma_{12}\sigma_{23}}{\sigma_{11}\sigma_{22} - \sigma_{12}^2} + b_{30}\frac{\sigma_{02}\sigma_{12} - \sigma_{01}\sigma_{22}}{\sigma_{11}\sigma_{22} - \sigma_{12}^2}$$

The second term on the right does not vanish, so that the OLS estimator from Model 1,

$$\hat{b}_{31}^F = \frac{m_{22}m_{13} - m_{12}m_{23}}{m_{11}m_{22} - m_{12}^2}$$

is, in fact, biased and the bias is given by the non-zero value of that second term.

This example shows that despite the specification error, OLS applied to Model 1 would have yielded an unbiased estimate for b_{32}, though not for b_{31}. One must, perhaps, be hungry indeed to take much nourishment from such a result. Still, one can imagine a realistic situation in which x_0, though unobserved, may (with reason) be postulated to behave according to Model 3. In that event, we can still do *something* with Model 1. Certainly, it is important to realize that specification error may have diverse effects, according to where it occurs. Hence, it is not enough for a skeptical critic to say, " The model is improperly specified, hence the estimates of structural coefficients are biased." He must, on the contrary—or *you* must, taking the role of critic—propose a new, " true " model that shows just wherein the initial model errs, and then compare the two to infer as much as possible about the consequences of the specification error.

Exercise. *We have seen that if Model 2 is true we can still salvage the x_3-equation from Model 1. If Model 3 is true, show that we can salvage the x_2-equation as well as an unbiased OLS estimate of b_{32}. If the true model is this,*

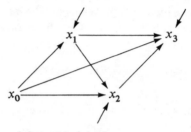

what can be salvaged from Model 1?

To illustrate still another situation, let the initial, possibly erroneous model be

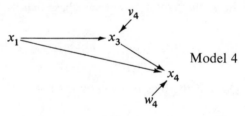

Model 4

Further consideration of relevant theory shows that the " true " model is

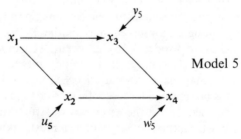

Model 5

This is an easy one.

Exercise.   *Show that the* OLS *estimator of* b_{43} *in Model 4 estimates* b_{43} *in Model 5 without bias and that the* OLS *estimator of* b_{41} *in Model 4 estimates* $b_{42}b_{21}$ *without bias.*

The principle is that insertion of an "intervening variable" into one path of an initial model does not invalidate that model, but merely elaborates it.

But suppose the "true" model is

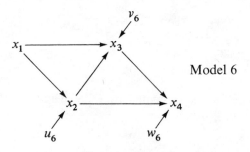

Model 6

The elements of this situation have already been covered in the episode of Model 1 versus Model 2, although the additional variable turns up at a different place in the causal ordering.

Exercise.   *If Model 6 is true, with the usual specification on its disturbances, show that the specification on the disturbances in Model 4 may include* $E(x_1 w_4) = E(x_1 v_4) = 0$ *but not* $E(x_3 w_4) = 0$. *Set forth the implications for* OLS *estimation of structural coefficients in Model 4.*

Here the principal is that insertion of an "intervening variable" into an initial model, where the intervening variable then operates as a common cause of two or more later variables, reveals specification errors in the initial model. Every such case must be evaluated on its own terms. In the illustration at hand (Model 6 versus Model 4) we may make reference to the earlier experience (Model 3 versus Model 1) to observe that specification error may have one-sided consequences. Let us write the "semireduced form" equations for Model 6, eliminating x_2 from the other two equations:

$$x_3 = (b_{31} + b_{32}b_{21})x_1 + b_{32}u_6 + v_6$$

$$x_4 = b_{43}x_3 + b_{42}b_{21}x_1 + b_{42}u_6 + w_6$$

Or, more compactly,

$$x_3 = a_{31}x_1 + v_6'$$

$$x_4 = a_{43}x_3 + a_{41}x_1 + w_6'$$

where

$$a_{31} = b_{31} + b_{32}b_{21}$$

$$a_{43} = b_{43}$$

$$a_{41} = b_{42}b_{21}$$

$$v_6' = b_{32}u_6 + v_6$$

$$w_6' = b_{42}u_v + w_6$$

We see that a_{31} may be estimated by OLS regression of x_3 on x_1, since $E(x_1 v_6') = 0$.

Exercise.   *Show that $E(x_1 v_6') = 0$.*

But OLS is not suitable for estimating a_{43} and a_{41}, since not all the explanatory variables are uncorrelated with w_6', the disturbance in the semireduced form of the x_4-equation. We note that the semireduced form of Model 6 is just Model 4, with the appropriate reservation concerning nonzero covariances of causal variables and disturbances.

Much theorizing in sociology takes the form of suggesting "intervening variables" to interpret causal linkages that have been recognized earlier. For the research worker using structural equation models, it becomes vital to know whether the hypotheses about these intervening variables tend to leave his work intact (though pointing toward a useful elaboration of it) or whether they tend to suggest that his estimates of causal influences are seriously biased. Since our examples of specification error have been nonnumerical, it is easy to miss the point that specification errors may engender biases that are real but nonetheless quantitatively trivial. In addition to a qualitative analysis of the nature of the bias (if any), the investigator will, therefore, often put hypothetical (but conceivable) values on certain structural coefficients in a "true" model in order to conjecture, with some plausibility, how wide of the mark he may be if he continues to work with the erroneous one. Such "sensitivity analysis" may sometimes provide comfort where otherwise the examination of specification error would yield pessimistic conclusions as to the acceptability of a model.

We have given only a few rudimentary illustrations of specification error but, it is hoped, enough to suggest that the serious investigator will muster all his ingenuity to anticipate threats to the validity of his results from this source. We have focussed on the "omitted variable" as a threat to validity, because this is one of the most common arguments encountered in discussions of models—perhaps because it is easy to suggest the name of a variable that has been overlooked, though not always so easy to justify a "true" model that includes it. Specification error arises in other ways, however.

Suppose one is in a position to compare estimates obtained for these two models:

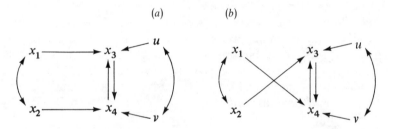

At least one of these two models ought to give results that are questionable in the light of theoretical expectations. If not, it would appear that the domain under study actually is not very highly structured in causal terms, or the measurements all are grossly contaminated by error, or our theory is virtually noncommittal on the significant issues. In any event, the clue to specification error in (a) or (b) is primarily the substantive implausibility of the estimates.

Exercise.   *Let the coefficients in diagram* (a) *be* b_{31}, b_{34}, b_{42}, b_{43} *and the coefficients in diagram* (b) *be* c_{32}, c_{34}, c_{41}, c_{43}. *Show that* $c_{43} = 1/b_{34}$ *and* $c_{34} = 1/b_{43}$. *Similarly, express* c_{32} *and* c_{41} *in terms of the b's. How would you answer a mathematician who claims that Models* (a) *and* (b) *are interchangeable, so that if one is "true" the other must be "true" too?*

In addition to wrongly omitted variables, the model may involve erroneously included variables. In theory, one such mistake should not be fatal if the model is otherwise correctly specified. The erroneously included variable should have a nearly zero coefficient. However, a mistake of this kind does impair the efficiency with which the model's coefficients are estimated. Hence, the investigator should not try to get by on the strategy of "including everything" on an initial run of his model. This strategy, pursued relentlessly, leads to underidentification, as we have seen in Chapter 6 (page 87).

Other issues properly subsumed under specification error include (*1*) tests of overidentifying restrictions (Chapter 3, pages 46–50; Chapter 7, pages 98–99); (*2*) the validity of the specification of linear and additive functional form, a matter that receives considerable attention in most statistical presentations of linear models; and (*3*) the acceptability of the homoskedasticity assumption in regard to disturbances, likewise treated in some statistics texts. The neglect of these last two topics in our sketch of this subject should not be mistaken for a judgement that they are unimportant. On the contrary, they are so important that they must be squarely faced throughout a project in constructing a model, but especially when considering the initial specification of the model. Statistical tests of linearity and homoskedasticity may be of use, but detailed inspection of the data aided by graphic plots is perhaps even more useful.

Exercise. *On page 17 we suggested that one could pose a countermodel to Model* II′ *as a means of discussing possible specification error in that model. On page 18, we made a similar suggestion regarding Model* III. *If you have not already done so, show how this might be done in each instance.*

FURTHER READING

The text by Rao and Miller (1971) is unusual in regard to the amount of attention given to matters relevant to this topic and also in that it is accessible to the reader not familiar with matrix algebra.

9

Measurement Error,
Unobserved Variables

From a formal point of view, the topic of error in (measurement of) variables is much the same thing as that of unobserved variables. All observation is fallible, no matter how refined the measuring instrument and no matter how careful the procedure of applying it. In a strict sense, therefore, we never measure exactly the true variables discussed in our theories. In this same strict sense, all (true) variables are "unobserved."

It may happen that errors of measurement are negligible, relative to the magnitudes of the disturbances in our equations or the standard errors of sampling in our estimates of coefficients. Of course, an investigator will not blithely assume this is so, but will make every effort to assess his measurement errors and their impact on his results. If the verdict is reassuring (perhaps because the other threats to valid inference are so uncomfortably large!) he may proceed, for the moment, to treat his variables as error-free, as we have been doing implicitly throughout this book.

A second possibility is that measurement error is appreciable but "well-behaved" and, perhaps, estimable. That is, a relatively simple and manageable specification of the behavior of the errors is acceptable. Either the errors can be shown not to impair seriously the results obtained, or their parameters can be estimated and corrections for their distortions can be introduced.

A third possibility—by all odds the most realistic one, in sociological investigations—is that our measuring instrument does not measure "cleanly" what we would like it to measure, but also reflects both random and systematic deviations from the "true" variable (the one discussed in the theory) which are both too large to neglect and too complicated ("messy") to manage on any elementary model of their behavior.

A counsel of perfection would be to work out a theory of errors and develop a model for the behavior of errors *pari passu* with the pursuit of the substantive objectives of every investigation. A more realistic kind of advice is predicated on the assumption that any single inquiry is just an incident in an historical stream of research. At a given moment, the investigator does all that he can to reduce errors of measurement, to estimate those that remain, and to accommodate his models and statistical procedures to the facts of measurement error, as best he can understand them on the basis of his own studies and the literature of his field. Over the long run, the cure for the more disruptive kinds of errors can only be improved techniques of measurement. But, as errors are tamed, it becomes possible to devise realistic and powerful theories of error, and to build directly into our models the requisite assumptions about errors. A mature science, with respect to the matter of errors in variables, is not one that measures its variables without error, for this is impossible. It is, rather, a science which properly manages its errors, controlling their magnitudes and correctly calculating their implications for substantive conclusions.

This is a very large topic. Indeed, almost the whole of psychometrics can be seen as a frontal assault on the problem of errors in variables. In sociology, substantial efforts to estimate error have been made in some survey research organizations. Moreover, sociologists have long been sensitized to the "indicator problem"—the often loose epistemic linkage between the variable specified in our theory and even the most plausible of the available empirical measures of it. Many chapters in the symposium edited by Goldberger and Duncan (1973) are concerned in one way or another with problems raised by models in which there are multiple indicators of unobserved variables and/or models in which measurement error complicates the estimation of structural coefficients. The symposium opens up several new, but difficult, approaches to these problems. We shall be content here only to suggest a few of the sorts of problems that arise as soon as one proposes to take explicit account of errors in variables.

Let us begin with this example.

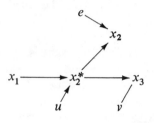

The primary causal model here is the simple causal chain, $x_1 \rightarrow x_2^* \rightarrow x_3$, already studied in Chapter 2 (with different notation). But now, one of the variables (x_2^*) is not directly observed. Instead, its observed counterpart (x_2) is contaminated with an error (e). Hence the complete model comprises three equations, one of which describes the fallible measurement of x_2^*, while the other two represent the causal model as such:

$$x_2 = x_2^* + e$$
$$x_2^* = b_{21} x_1 + u$$
$$x_3 = b_{32} x_2^* + v$$

To make the example easy, we require the error of measurement to be well-behaved. (Since we are talking about the properties of the model and not about what the world is really like, it will be recognized that this requirement is itself problematic in any realistic context.) Specifically, we assume that e is uncorrelated with x_2^* and also with both x_1 and x_3:

$$E(x_2^* e) = E(x_1 e) = E(x_3 e) = 0.$$

That is, roughly speaking, the error is "random" and not "systematic." Moreover, $E(e) = 0$, just as for all other variables in the model. We employ the usual specification in regard to disturbances:

$$E(x_1 u) = E(x_2^* v) = E(x_1 v) = 0.$$

In consequence of these specifications, we find that $E(eu) = E(ev) = E(uv) = 0$.

Exercise.   *Verify that $E(eu) = E(ev) = E(uv) = 0$.*

We find, further, upon squaring x_2 and taking expectations, that

$$\sigma_{22} = \sigma_{22}^* + \sigma_{ee}$$

using the symbol σ_{22}^* for $E[(x_2^*)^2]$. That is, the total variance of the fallible measurement (x_2) is the sum of the variances of the true but unobserved variable we are trying to measure (x_2^*) and the variance of our measurement errors (e).

Another useful and perhaps surprising result is that the covariances are not affected by the error in measurement. That is, $E(x_1 x_2) = E(x_1 x_2^*)$ and $E(x_2 x_3) = E(x_2^* x_3)$.

Exercise.   *Verify this.*

Thus, when we multiply through the x_2^*-equation by x_1 we obtain

$$\sigma_{12} = b_{21}\sigma_{11}$$

so that

$$b_{21} = \frac{\sigma_{12}}{\sigma_{11}}$$

We see that the OLS estimator m_{12}/m_{11} estimates b_{21} without bias. In this segment of the model, the well-behaved measurement error has a relatively benign effect. The effect is not wholly benign, however. We must reckon with its influence on the precision of our estimator. Substituting the x_2^*-equation into the equation relating the observed to the true value, we obtain

$$x_2 = b_{21}x_1 + u + e$$

Hence the total variance of x_2 is

$$\sigma_{22} = b_{21}^2 \sigma_{11} + \sigma_{uu} + \sigma_{ee}$$

(obtained by squaring both sides and taking advantage of the vanishing covariances noted earlier). Regarding the "explained" variance,

$b_{21}^2 \sigma_{11}$, as fixed, we note that the "unexplained" variance will increase as σ_{ee} increases. Thus, greater error variance means a larger standard error for our OLS estimate of b_{21}. This effect can, of course, be offset by drawing a larger sample.

So much for the good news. Now for the bad news. We substitute for x_2^* in the x_3-equation the expression $x_2^* = x_2 - e$:

$$x_3 = b_{32} x_2 + v - b_{32} e$$

or

$$x_3 = b_{32} x_2 + v'$$

where $v' = v - b_{32} e$.

We are tempted to regress x_3 on x_2, that is, to use the OLS estimator of b_{32}, m_{23}/m_{22}. But, upon multiplying the x_3-equation through by x_2 we obtain

$$\sigma_{23} = b_{32} \sigma_{22} + E(x_2 v')$$

$$= b_{32} \sigma_{22} - b_{32} E(x_2 e)$$

$$= b_{32}(\sigma_{22} - \sigma_{ee})$$

whence

$$b_{32} = \frac{\sigma_{23}}{\sigma_{22} - \sigma_{ee}}$$

Therefore, our OLS procedure estimates *not* b_{32} but rather σ_{23}/σ_{22}, or

$$b_{32}\left(1 - \frac{\sigma_{ee}}{\sigma_{22}}\right) = b_{32} \frac{\sigma_{22}^*}{\sigma_{22}^* + \sigma_{ee}}$$

Hence, the greater the variance in the errors (that is, the lower the precision of our measurements) the greater the downward bias in our OLS estimate of b_{32}. (The bias actually is downward only if b_{32} is positive. The bias is toward zero, regardless of the sign of b_{32}.)

To generalize: Error in the dependent variable, if "well behaved," does not bias the OLS estimate. But error in the independent variable, even though "well behaved," imparts a downward bias to the OLS estimate. The kinship between "measurement error" and "specification error" is brought out by this last result. Our difficulty with OLS arose in writing the x_3-equation in terms of the observed variable x_2 and the disturbance v', because, in this formulation, the explanatory variable is not uncorrelated with the disturbance, that is, $E(x_2 v') \neq 0$.

But let us return to the good news. Adopting a different approach to the estimation of b_{32}, we multiply through the x_3-equation

$$x_3 = b_{32} x_2 + v'$$

by x_1 (rather than x_2) to obtain

$$\sigma_{13} = b_{32} \sigma_{12}$$

inasmuch as $E(x_1 v') = E(x_1 v) - b_{32} E(x_1 e) = 0$. Finding that $b_{32} = \sigma_{13}/\sigma_{12}$, we are led to propose instead of the OLS estimator of b_{32} (to wit, m_{23}/m_{22}) the IV estimator, m_{13}/m_{12}, using x_1 as an instrument for the contaminated x_2. But from the original formulation of the x_3-equation

$$x_3 = b_{32} x_2^* + v$$

we find (as previously noted) that

$$\sigma_{23} = b_{32} \sigma_{22}^* = b_{32}(\sigma_{22} - \sigma_{ee})$$

recalling that $E(x_2 x_3) = E(x_2^* x_3)$. Now, we have the IV estimate $\hat{b}_{32} = m_{13}/m_{12}$ and we can estimate σ_{23} and σ_{22} directly from our sample data. From these, we derive an estimate of the error variance,

$$\hat{\sigma}_{ee} = \frac{\hat{b}_{32} \hat{\sigma}_{22} - \hat{\sigma}_{23}}{\hat{b}_{32}}$$

It is, therefore, a remarkable property of this model that one can not only secure unbiased estimates of its coefficients, despite the presence of measurement error, but actually estimate the crucial parameter of the measurement process itself. Clearly, this remarkable property is due to the fact that the three-variable, simple causal chain model is overidentified to begin with. Thus, we infer the principle that overidentification provides a possible weapon in an attack on measurement error. But the weapon is no better than the theory of error that provides its ammunition. (Our model, it should be clear, rests on a very strong theory of error.) Moreover, the overidentifying restriction(s) must be in the "right" place, relative to the location of the measurement error, in order for the weapon to work at all. To appreciate this last point, carry out this exercise.

Exercise.   *Suppose the fallible variable is x_1 in a simple causal chain model, and that the error is "well-behaved" as before:*

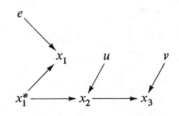

Show that an unbiased estimate of b_{21} cannot be obtained from sample moments, whether by OLS or IV, despite the availability of an overidentifying restriction on the model.

The exercise prepares the ground for another example, a model in which we dispense with the overidentifying restriction:

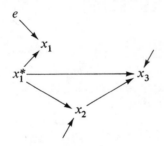

We already know (from the exercise) the nature of the difficulty with the x_2-equation. Let us, therefore, focus on the x_3-equation or, equivalently, the model

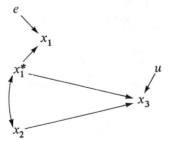

Continuing with the now-familiar specifications on the error and disturbance terms, we have $E(x_1^* e) = E(x_2 e) = E(x_3 e) = E(x_1^* u) = E(x_2 u) = 0$, from which it follows that $E(x_1 u) = E(eu) = 0$. The equations of the model are

$$x_1 = x_1^* + e$$

$$x_3 = b_{31} x_1^* + b_{32} x_2 + u$$

We note that $\sigma_{11} = \sigma_{11}^* + \sigma_{ee}$, and that $E(x_1^* x_2) = \sigma_{12}$ and $E(x_1^* x_3) = \sigma_{13}$. We find (verify this) that

$$\sigma_{13} = b_{31}(\sigma_{11} - \sigma_{ee}) + b_{32}\sigma_{12}$$

$$\sigma_{23} = b_{31}\sigma_{12} + b_{32}\sigma_{22}$$

Hence,

$$b_{31} = \frac{\sigma_{13}\sigma_{22} - \sigma_{12}\sigma_{23}}{\sigma_{11}\sigma_{22} - \sigma_{12}^2 - \sigma_{ee}\sigma_{22}}$$

$$b_{32} = \frac{\sigma_{11}\sigma_{23} - \sigma_{12}\sigma_{13} - \sigma_{ee}\sigma_{23}}{\sigma_{11}\sigma_{22} - \sigma_{12}^2 - \sigma_{ee}\sigma_{22}}$$

From the presence of terms involving σ_{ee} in the denominator of b_{31} and both the numerator and denominator of b_{32}, we infer that the OLS estimates obtained in regressing x_3 on x_2 and x_1 will not be satisfactory in regard to either structural coefficient. The formula (above) for b_{31} takes the form $K_1/(K_2 - K_3)$, where K_1 is the numerator, $K_2 = \sigma_{11}\sigma_{22} - \sigma_{12}^2$, and $K_3 = \sigma_{ee}\sigma_{22}$. Both K_2 and K_3 are intrinsically positive, although K_1 may be either positive or negative. The OLS procedure estimates not b_{31} but K_1/K_2. Therefore, we conclude that there is a downward bias in the absolute value of the OLS estimate. To see the bias in the OLS estimate of b_{32}, we derive another formula for that coefficient:

$$b_{32} = \frac{\sigma_{11}\sigma_{23} - \sigma_{12}\sigma_{13}}{\sigma_{11}\sigma_{22} - \sigma_{12}^2} - \frac{b_{31}\sigma_{12}\sigma_{ee}}{\sigma_{11}\sigma_{22} - \sigma_{12}^2}$$

Exercise.   *Verify this.*

By OLS, we estimate only the first term, or

$$b_{32} + \frac{b_{31}\sigma_{12}\sigma_{ee}}{\sigma_{11}\sigma_{22} - \sigma_{12}^2}$$

The sign of the bias depends only on the signs of b_{31} and σ_{12}. If both of these are positive, we overestimate b_{32} (while underestimating b_{31}).

This situation is hopeless, unless we can secure auxiliary information about σ_{ee}. Various experimental designs for estimating σ_{ee} are available. The choice of an appropriate one depends (among other things) on the nature of the variables, and the literature of a well-developed substantive field will usually include studies specifically designed to estimate measurement error. Suppose that the true variable is considered to be a relatively permanent attribute for each member of the population and suppose that we can make a (fallible) measurement of it more than once without any carry over of error from one occasion to another. (This assumption clearly is implausible if memory or learning is involved or if the very act of measurement releases causal forces that otherwise would not come into the picture.) If the true variable is y^*, a model for the experiment of carrying out two measurements on a sample of members of the population is

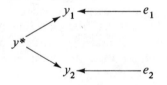

or, more explicitly, $y_1 = y^* + e_1$ and $y_2 = y^* + e_2$.

On each occasion, according to the model, the error is uncorrelated with the true value, so that $E(y^*e_1) = E(y^*e_2) = 0$. By ruling out "carry over" we mean, more precisely, to specify $E(e_1 e_2) = 0$. The supposition of no systematic error is conveyed by $E(e_1) = E(e_2) = 0$. If the experiment is well executed, so that each measurement simulates faithfully the conditions encountered in a substantive investigation, then the variance of measurement errors should be the same: $E(e_1^2) = E(e_2^2)$. This is one assumption of the experiment that can actually be tested, provided, of course, that one accepts the prior assumption that y^* itself is unchanged between occasions. Our model implies that

$$E(y_1^2) = \sigma_{yy}^* + E(e_1^2)$$

and

$$E(y_2^2) = \sigma_{yy}^* + E(e_2^2)$$

Hence, if $E(e_1^2) = E(e_2^2)$, $E(y_1^2) = E(y_2^2)$. Statistical procedures for testing for homogeneity of variances are available. If the data are consistent with the hypothesis of homogeneity, that is, the null hypothesis $E(y_1^2) = E(y_2^2)$, we may derive an estimate of the error variance from the model by noting, first, that the model implies

$$y_1 - y_2 = e_1 - e_2$$

whence

$$E(y_1^2) + E(y_2^2) - 2E(y_1 y_2) = E(e_1^2) + E(e_2^2)$$

since $E(e_1 e_2) = 0$ but $E(y_1 y_2) \neq 0$. But if $E(y_1^2) = E(y_2^2) = \sigma_{yy}$ and $E(e_1^2) = E(e_2^2) = \sigma_{ee}$ (one will have decided already to accept this assumption before proceeding), we have

$$\sigma_{ee} = \sigma_{yy} - \sigma_{12}$$

The variance σ_{yy} is estimated upon pooling m_{11} and m_{22}, and the covariance σ_{12} is estimated from m_{12}.

Given the estimate of σ_{ee}, we may return to the formulas already given for b_{31} and b_{32} and see at once that this is the one item of information needed to "correct" the OLS estimates for their biases.

In our final example of a model taking account of "well-behaved" errors of measurement, we raise the question of whether principles observed to operate in recursive models carry over to the nonrecursive case. The model, a slightly modified version of one discussed by Goldberger (1972), is represented by the following path diagram:

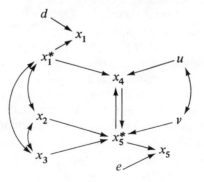

The structural equations are

$$x_4 = b_{41} x_1^* + b_{45} x_5^* + u$$
$$x_5^* = b_{52} x_2 + b_{53} x_3 + b_{54} x_4 + v$$

The usual specification on the disturbance terms is adopted:

$$E(x_1^* u) = E(x_1^* v) = E(x_2 u) = E(x_2 v) = E(x_3 u) = E(x_3 v) = 0.$$

The equations describing the behavior of measurement errors are

$$x_1 = x_1^* + d$$
$$x_5 = x_5^* + e$$

To make the errors "well-behaved," we specify

$$E(x_1^* d) = E(x_2 d) = E(x_3 d) = E(x_4 d) = E(x_5^* d)$$
$$= E(x_1^* e) = E(x_2 e) = E(x_3 e) = E(x_4 e) = E(x_5^* e) = 0.$$

We also have to make explicit the specification, implied by the diagram, that $E(de) = 0$. (The assumption that errors in two variables are uncorrelated will often be a problematic one in a realistic application.) The specifications already stated suffice to guarantee that measurement errors are uncorrelated with structural disturbances.

Exercise.   *Show that* $E(du) = E(dv) = E(eu) = E(ev) = 0.$

But note that nothing that has been said would imply that the disturbances are uncorrelated; we must suppose, in general, that $E(uv) \neq 0$.

We recall from an earlier example that measurement error in an exogenous variable causes trouble. But in that example the method of estimation under consideration was OLS. Here, we already know that OLS is not suitable, so we have to reopen the question of whether estimates will be biased by the error in measuring x_1^*. We also recall earlier examples in which measurement error in the dependent variable did not bias OLS estimates. But now, x_5^* is not only a dependent variable, but is simultaneously a causal variable (with respect to x_4). Again, we cannot infer without special study of the issue that the principle developed earlier will carry over to the nonrecursive case.

Let us first examine the x_5^*-equation. We can immediately restate it as an x_5-equation by substituting its right-hand side for x_5^* in $x_5 = x_5^* + e$. This yields

$$x_5 = b_{52}x_2 + b_{53}x_3 + b_{54}x_4 + v'$$

where $v' = e + v$. We see that this equation is just identified. There are three coefficients to estimate, and three (observed) instrumental variables are available, to wit, x_1, x_2, and x_3. Their eligibility as instrumental variables follows from the fact that each is uncorrelated with the new disturbance, v', since each is uncorrelated with both the original disturbance (v) and the measurement error (e).

Query: How do we know, in particular, that $E(x_1 v) = 0$ and $E(x_1 e) = 0$, since these are not explicit specifications of the model?

We may therefore proceed to IV estimation of the rewritten x_5-equation. Writing out the covariances in the usual fashion and substituting corresponding sample moments for them, we find

$$m_{15} = m_{12}\hat{b}_{52} + m_{13}\hat{b}_{53} + m_{14}\hat{b}_{54}$$
$$m_{25} = m_{22}\hat{b}_{52} + m_{23}\hat{b}_{53} + m_{24}\hat{b}_{54}$$
$$m_{35} = m_{23}\hat{b}_{52} + m_{33}\hat{b}_{53} + m_{34}\hat{b}_{54}$$

so that we may solve for the $\hat{b}$'s by any convenient algorithm.

We turn our attention to the x_4-equation and note that we can eliminate the unobserved variables from it by the substitutions, $x_1^* = x_1 - d$ and $x_5^* = x_5 - e$, to obtain

$$x_4 = b_{41}x_1 + b_{45}x_5 + u'$$

where $u' = u - b_{41}d - b_{45}e$. We see that neither x_1 nor x_5 can serve as an instrumental variable, since $E(x_1 u') \neq 0$ and $E(x_5 u') \neq 0$.

Exercise. *Verify these results.*

However, since $E(x_2 u') = E(x_3 u') = 0$, we have two instrumental variables available. Since there are only two structural coefficients to estimate, the rewritten x_4-equation is just identified.

Exercise. *Verify that $E(x_2 u') = E(x_3 u') = 0$.*

IV estimates are the solutions of the following pair of equations:

$$m_{24} = m_{12}\hat{b}_{41} + m_{25}\hat{b}_{45}$$
$$m_{34} = m_{13}\hat{b}_{41} + m_{35}\hat{b}_{45}$$

To consolidate our results: Well-behaved measurement error in an endogenous variable does not render the IV method (or the two-stage regression procedure mentioned in the next chapter, where there are "too many" instrumental variables) unsuitable for estimating an equation in a nonrecursive model. (We note, without further discussion, that it does inflate standard errors of estimated coefficients, however.) Moreover, well-behaved measurement error in an exogenous variable does not rule out estimation by use of instrumental variables provided that the original structural equation is overidentified. If there are enough predetermined variables in the model to provide a sufficient number of instrumental variables (at least as many as the number of coefficients to be estimated), we may set up estimating equations, on the IV principle, whose solution will yield the desired estimates.

Two noteworthy features of this example may be mentioned. First, it is not sufficient for the model to have an overidentified equation "somewhere." The overidentifying restriction(s) must be in the "right place." In a complicated model, it may require extensive analysis to be sure whether this holds true. Second, it is of interest that, although x_1 was eligible as an instrument in estimating the x_5-equation, it was no longer so for the x_4-equation. Again, careful analysis is required to take advantage of such subtle properties of the model.

Finally, we note that observations on the five variables in this model disclose nothing about the magnitude of the errors in x_5. Without additional evidence, we cannot estimate σ_{ee}. For x_1, on the contrary, the model itself provides a method of estimating the error variance. Recall the "solved-out" version of the x_4-equation:

$$x_4 = b_{41}x_1 + b_{45}x_5 + u - b_{41}d - b_{45}e$$

We multiply through by x_1 and take expectations:

$$\sigma_{14} = b_{41}\sigma_{11} + b_{45}\sigma_{15} - b_{41}\sigma_{dd}$$

Exercise. *You should verify that* $E(x_1u) = E(x_1e) = 0$ *and* $E(x_1 d) = \sigma_{dd}$, *since these facts have not hitherto been made explicit.*

We already have estimates of the b's, and the σ's can be estimated directly from the sample. After substituting the estimates for the parameters in this equation, we may solve for $\hat{\sigma}_{dd}$.

Exercise. *Show that the same approach will not enable us to estimate* σ_{ee}.

Our illustrations in this chapter have exemplified two broad strategies for coping with measurement error, if its location gives rise to difficulties in estimation:

(*1*) Imbed a model of measurement error in a substantive model that is otherwise overidentified and use the overidentifying restriction(s) as a means of estimating error variance (or covariance, if that, too, is present) and estimating structural coefficients free of bias due to measurement error.

(*2*) Use a model of measurement error to conduct an auxiliary investigation designed to estimate error variance(s) with which to "correct" the estimates of structural coefficients.

The two strategies are not mutually exclusive, but may be used in tandem. Variants of each are conceivable. If, in an auxiliary experiment, one can actually measure $y^* = y_1$ directly (that is, with negligible error) at great pains and expense and at the same time obtain y_2 by the standard procedure, then σ_{ee} is given directly by $E(y_2^2) - E(y_1^2)$ and both these quantities are estimated from the experimental data. (Note that a minimal test of the model is that $\hat{\sigma}_{22} > \hat{\sigma}_{11}$.)

A variant of the first strategy arises in connection with multiple indicators of an unobserved variable, each of which is not only fallible (perhaps highly so) but also potentially subject to systematic distortions. Such a situation demands a very subtle but nonetheless powerful theory concerning sources of error in measurement. Standard multivariate procedures for coping with a plethora of indicators (factor analysis, principal component analysis, canonical correlation, for example) are available, but they may not always prove satisfactory for complicated sociological problems. In the next chapter we try to suggest the character of the complications.

Another especially ugly complication is that many sociological variables can take on only a small number of discrete values and are subject to hard and fast floor and/or ceiling effects. Suppose the true variable is the number of rooms in an apartment or house. The minimum is one and the maximum, though not well defined, may be, effectively, seven or eight. There can be no negative error if $y^* = 1$ and (almost) no positive error if $y^* = 8$. Under such circumstances to assume that $E(y^*e) < 0$ may be more plausible than to specify $E(y^*e) = 0$. Situations like this—no doubt there are numerous significantly different variants thereof—require more systematic investigation than they have received. We are only beginning to realize that measurement problems require theories quite as powerful and procedures just as well controlled as those needed to execute a study not fatally flawed by sampling error and specification error.

FURTHER READING

Although the psychometric literature contains much material on measurement error, it is less useful than it might be for present purposes, since the approach is usually via correlation and standardized variables. See Walker and Lev (1953, Chap. 12) for an introduction to this literature. In the econometric literature, Wonnacott and Wonnacott (1970, Chap. 7) is especially instructive in that error in the independent variable is considered in the same context as other sources of correlation between regressor and disturbance. Two highly recommended papers presenting models involving measurement error are D. E. Wiley and J. A. Wiley (Chap. 21 in Blalock, 1971) and D. E. Wiley (Chap. 4 in Goldberger and Duncan, 1973).

10

Multiple Indicators

Terminology to designate unobserved variables tends to vary by discipline. In psychometrics, where one fallible score is at stake, the unobserved counterpart is called a "true score"; errors of measurement give rise to "unreliability"; and using an estimate of reliability to correct the estimates of structural coefficients is termed "correcting for attenuation." Where there are several (in practice, four or more) fallible measurements for each of one or more unobserved variables, the latter are conceptualized as "factors," and the theory of factor analysis is brought into play.

In economics, recognition of measurement error (random or systematic) is often symbolized by referring to observed variables as "proxy" variables, acknowledging that they are not identical with the (unobserved) variables discussed in economic theory. The examples in Chapter 9 disclose some ways in which estimates of structural coefficients may be distorted by naively replacing the unobserved variable in an equation by its "proxy." Economists are also sensitized to the possibility of hypothetical constructs like "permanent income," for which actual measurements of income (even if free of observational error) afford only a proxy, owing to the random variability introduced into actual money receipts from time to time by transitory causes.

In sociology, this generic problem often is referred to as the problem of validity of "indicators." Perhaps there is no direct measure of such a concept as "social cohesion," for example, but one theory of the phenomenon would suggest that the suicide rate is an "indicator" of social cohesion. There are many reasons, however, why the relationship between concept and indicator is contingent and loose rather than fixed and exact. In view of these reasons (we will not spell them out here), it is sometimes felt that the investigator is better off with multiple indicators, on the theory (we surmise) that there is then a chance that an error in one indicator may be offset by a compensating error in another.

Here, we can only illustrate a few of the issues that arise when an investigator resolves to take seriously the problem of indicator validity. Our premise is that there is no *general* solution to this problem, but that it has to be attacked on its merits in connection with each substantive model in which indicator variables appear. The reader is warned at the outset that a major purpose of this chapter is to provide motivation for further study. Hence, we repeatedly note problems in estimation, optimal solutions to which require resort to methods beyond the scope of this volume.

Let us modify slightly the last example in Chapter 9:

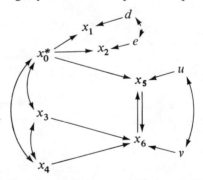

The new feature is that there are now two distinct indicators (or fallible measurements) of the unobserved variable. Hence, we have the two equations,

$$x_1 = x_0^* + d$$

$$x_2 = ax_0^* + e$$

Note that one of these equations must include a "scale factor" a since, in general, we do not require that x_1 and x_2 be measured on the same scale. (Suicide rate and per capita contributions to social welfare agencies might be two such noncommensurable indicators of "social cohesion"; but the reader must henceforth supply his own illustrative names for variables.) It is simply a matter of convenience which of the two indicators we designate as x_1 (without a scale factor). The consequence of this choice is that we thereby establish an implicit metric for the unobserved variable. We require the errors, d and e, to be "well behaved," so that they are uncorrelated with x_0^* and also with observed variables $x_3, \ldots, x_6$. As usual, predetermined variables are uncorrelated with structural disturbances. Thus, x_0^*, x_3, and x_4 are all uncorrelated with both u and v (although the latter may well have a nonzero covariance). These specifications suffice to insure that both x_1 and x_2 are uncorrelated with structural disturbances u and v.

Exercise.   *Verify this, if it is not already "obvious."*

We leave open for later discussion the question of whether $E(de) = 0$, and the diagram shows a dashed curve connecting the errors d and e. The structural equations are

$$x_5 = b_{50}x_0^* + b_{56}x_6 + u$$

$$x_6 = b_{63}x_3 + b_{64}x_4 + b_{65}x_5 + v$$

We note that the status of each of these equations with respect to identification is made problematic by the presence in the model of the unobserved variable and the indicator variables.

A general heuristic device for studying models containing unobserved variables is to "solve out" such variables, thereby deriving new equations. These are then examined in regard to identification and problems of estimation. Let us substitute $x_0^* = x_1 - d$ (from the x_1-equation) for x_0^* in the x_2-equation:

$$x_2 = ax_1 + e' \quad \text{where} \quad e' = e - ad$$

We see that four of the variables in the model, $x_3, \ldots, x_6$, are eligible as instruments, since each is uncorrelated with both d and e and, therefore, with e'. At first glance it may seem odd that the list of instrumental variables includes the two endogenous variables, x_5 and x_6. A closer look reveals that the model is, in effect, block recursive, with the x_5- and x_6-equations comprising one block, the x_1- and x_2-equations the other. With this embarrassment of riches, the rewritten x_2-equation—or, precisely, its one coefficient, a—is clearly overidentified. We are tempted to proceed at once to estimation, resolving the problem of an excess number of instrumental variables in a manner reminiscent of 2SLS. That is, we would regress x_1 on x_3, x_4, x_5, and x_6 to calculate $\hat{x}_1$ and then estimate a from the OLS regression of x_2 on $\hat{x}_1$. OLS is legitimate at this stage, since $E(\hat{x}_1 e') \cong 0$, given that $\hat{x}_1$ is a linear combination of $x_3, \ldots, x_6$, each of which is uncorrelated with e'.

Before beginning this calculation, however, we should note that our new equation can also be written,

$$x_1 = \frac{1}{a} x_2 + d'$$

where

$$d' = d - \frac{e}{a}$$

Again, we have an overidentified equation and, again, it would perhaps seem that a two-stage regression procedure would provide a straightforward approach. Moreover, it hardly matters whether we estimate a or $1/a$; knowledge of one is equivalent to knowledge of the other, in principle. The unfortunate fact is, however, that such a procedure is sensitive to the choice of normalization rule. That is, if the estimate $\hat{a}$ is obtained from the solved-out x_2-equation and $\widehat{1/a}$ from the solved-out x_1-equation (using two-stage regression with the same list of instrumental variables), then $1/\hat{a} \neq \widehat{1/a}$ in general. Moreover, the regression results provide no basis for preferring one of these estimates to the other or for averaging or otherwise reconciling them. The awkward situation is that this estimation problem does not yield nicely to the seemingly straightforward two-stage regression approach. If the investigator finds this too disconcerting, he must resort to more advanced methods—too advanced to fall within the scope of this book.

[Not only the two-stage regression procedure mentioned here, but also 2SLS itself, in the standard simultaneous-equation context, is sensitive to the normalization rule. We did not mention this in Chapter 7, on the assumption that the investigator will have resolved the question of normalization before proceeding to estimation. Fisher (reprinted in Blalock, 1971) remarks: "This is not an unreasonable assumption, as such normalization rules are generally present in model building, each variable of the model being naturally associated with that particular endogenous variable which is determined by the decision-makers whose behavior is represented by the equation [page 264]." Whether or not Fisher's view is plausible, there clearly is no "natural" normalization rule in the present instance.]

There is another instructive way to look at the overidentification of the x_1- and x_2-equations. Let us multiply through each of them by all the eligible instrumental variables; we find

$$\sigma_{13} = \sigma_{03}^* \qquad \sigma_{23} = a\sigma_{03}^*$$

$$\sigma_{14} = \sigma_{04}^* \qquad \sigma_{24} = a\sigma_{04}^*$$

$$\sigma_{15} = \sigma_{05}^* \qquad \sigma_{25} = a\sigma_{05}^*$$

$$\sigma_{16} = \sigma_{06}^* \qquad \sigma_{26} = a\sigma_{06}^*$$

The availability of several distinct ways of calculating the parameter a means that overidentifying restrictions are implied in setting those solutions equal to each other:

$$a = \frac{\sigma_{23}}{\sigma_{13}} = \frac{\sigma_{24}}{\sigma_{14}} = \frac{\sigma_{25}}{\sigma_{15}} = \frac{\sigma_{26}}{\sigma_{16}}$$

Although these equalities must hold in the population (if the model is true), they can be expected to hold at best only approximately in the sample. But if there are marked departures from equality in the sample data (too large to attribute to sampling error), we have evidence of some sort of specification error. Later examples will suggest some ways in which this could come about. But it should be noted—you have presumably come to expect this kind of "bad news"—that rejection of the overidentifying restrictions (or even some particular subset of them) does not provide unambiguous evidence concerning the nature and location of the specification error. At best, this outcome provides clues to the specification error, and such clues must be interpreted with great care and caution (Costner & Schoenberg, 1973).

Let us turn our attention to the other two equations, which, after all, comprise the substance of the model. Looking first at the x_6-equation, which has three explanatory variables, we find that there are actually four variables available as instruments: $x_1, \ldots, x_4$, since each of them is uncorrelated with the disturbance v. (At this juncture, by the way, it does not matter whether $\sigma_{de} = 0$.) Hence, the x_6-equation is overidentified, and it seems reasonable to proceed with two-stage estimation, regressing x_5 on x_1, x_2, x_3, and x_4 to calculate $\hat{x}_5$ and then regressing x_6 on x_3, x_4, and $\hat{x}_5$. It is a comfort to know that at least this much of the model is relatively unproblematic. However, if there is any concern about the correctness of our specifications on the x_1- and x_2-equations, this had best be resolved first, because any respecification of these equations might affect the eligibility of these variables as instruments.

To estimate the x_5-equation, we will need to "solve out" the unobserved variable. For example, using the x_1-equation for this purpose we obtain

$$x_5 = b_{50}x_1 + b_{56}x_6 + u'$$

where

$$u' = u - b_{50}d$$

It appears that the equation is just identified, since only x_3 and x_4 are available as instrumental variables, if we wish to keep open the possibility that $\sigma_{de} \neq 0$. On the other hand, if $\sigma_{de} = 0$, then x_2 as well, though not x_1 (*Why?*), can serve as an instrument. Working with just x_3 and x_4, IV estimates of the structural coefficients may be obtained as solutions to

$$m_{35} = \hat{b}_{50}m_{13} + \hat{b}_{56}m_{36}$$
$$m_{45} = \hat{b}_{50}m_{14} + \hat{b}_{56}m_{46}$$

But our conclusion was too hasty, for we could equally well have used the x_2-equation to "solve out" x_0^* from the x_5-equation:

$$x_5 = (b_{50}/a)x_2 + b_{56}x_6 + u''$$

where

$$u'' = u - b_{50}e/a$$

and IV estimates would be

$$m_{35} = \widehat{(b_{50}/a)}m_{23} + \hat{b}_{56}m_{36}$$

$$m_{45} = \widehat{(b_{50}/a)}m_{24} + \hat{b}_{56}m_{46}$$

We would then obtain $\hat{b}_{50}$ making use of our previously obtained estimate, $\hat{a}$ (supposing that little problem to have been solved!). The two sets of IV estimates would not, in general, be the same, although they should be "approximately" so if the model is true. We must conclude that the x_5-equation is overidentified, but this comes about in such a way that two-stage regression does not provide a way to resolve the problem.

Indeed, the problem is not merely that of deciding which one of these two estimators to accept. We can "solve out" x_0^* from the x_5-equation using not only the x_1-equation or the x_2-equation, but any weighted combination of the two equations. Suppose W is a constant. We may write

$$Wx_1 = Wx_0^* + Wd$$

$$(1 - W)x_2 = (1 - W)ax_0^* + (1 - W)e$$

whence

$$x_0^* = \frac{Wx_1 + (1 - W)x_2 - Wd - (1 - W)e}{W + (1 - W)a}$$

This solution, for whatever value of W, may be used to "solve out" x_0^* from the x_5-equation. (If $W = 1$, we are using, in effect, the x_1-equation alone, as before; if $W = 0$, we are using the x_2-equation alone.) Once this is done—supposing a value of W and an estimate of a are available to substitute into the formula—we can proceed to IV estimation of the x_5-equation using x_3 and x_4 as instrumental variables.

But the calculation of W is our new stumbling block, if we insist on seeking a statistically efficient method of estimation, rather than one that merely tends to be unbiased in large samples. Clearly, W should be chosen in such a way that the sampling variance of our estimates, $\hat{b}_{50}$ and $\hat{b}_{56}$, is a minimum. The solution to this difficult problem is, again, beyond our scope. (Actually, this formulation of the issue, in terms of choosing an optimal value of W, is introduced only to illustrate the complexity of the estimation problem that arises with multiple indicators. We do not mean to suggest that calculation of W will be an explicit step in an efficient method of estimation. On the contrary, such a method will, in effect, accomplish the requisite weighting at the same time that it produces the appropriate estimate of the overidentified parameter, a. This same remark applies at later points in the chapter, where similar issues of "weighting" arise.)

One final aspect of this model to be considered arises in evaluating the variances and covariance of our indicator variables. We find, by the usual technique, that

$$\sigma_{11} = \sigma_{00}^* + \sigma_{dd}$$

$$\sigma_{22} = a^2 \sigma_{00}^* + \sigma_{ee}$$

$$\sigma_{12} = a\sigma_{00}^* + \sigma_{de}$$

We see that if σ_{de} is specified to be zero, the parameters σ_{00}^*, σ_{dd}, and σ_{ee} are just identified; for in that event the three equations can be solved for the three unknowns. Assuming we have already estimated a, we could use the sample data to compute

$$\hat{\sigma}_{00}^* = \frac{\hat{\sigma}_{12}}{\hat{a}}$$

$$\hat{\sigma}_{dd} = \hat{\sigma}_{11} - \hat{\sigma}_{00}^*$$

$$\hat{\sigma}_{ee} = \hat{\sigma}_{22} - \hat{a}^2 \hat{\sigma}_{00}^*$$

This operation is something of a side issue, in that it contributes nothing to estimation of coefficients in our two main equations. It may be of some use, however, in appraising our indicators; for we can now examine the values of $\hat{\sigma}_{dd}/\hat{\sigma}_{11}$ and $\hat{\sigma}_{ee}/\hat{\sigma}_{22}$ as indices of unreliability. In future research, if forced to choose between the indicators, we would presumably prefer the one with the lesser value of this index. But in using this criterion, we should want to be rather sure of the specification $\sigma_{de} = 0$, for there is no way to test it with data on our six observed variables; nor does the value of σ_{de} (if it is not zero) have any other direct consequence for our procedures.

Summing up, what has been gained by the use of the two indicators? (We seemed to get along perfectly well with only one in the otherwise similar model in Chapter 9.) First, in view of the overidentifying restrictions on the parameter a, we have an opportunity to test whether our indicators really are "well behaved," as the model assumes them to be. If the truth of the overidentifying restrictions is not called into question by the result of a suitable test, we are somewhat reassured on this score. If the outcome is otherwise, we know that one or both indicators are "contaminated" in a way that the model fails to take into account. We must reconsider the model or the indicators.

Second, assuming we continue to accept the overidentifying restrictions after testing them, we have a clue as to which indicator is more reliable, if that information is of any use, and if we are willing to make the strong assumption that the indicators are correlated only to the extent of their common dependence on the unobserved variable.

Third, again on the assumption that the overidentifying restrictions are acceptable, the sampling errors of our estimates of structural coefficients ordinarily will be smaller when using two (well-behaved) indicators, rather than either of them alone.

Despite these gains, the contribution of multiple indicators should not be exaggerated. Their help in connection with the identification problem may be one-sided. We have seen that each of the indicators (x_1 and x_2) of the unobserved variable is eligible as an instrument in estimating the equation in which the unobserved variable x_0^* does *not* appear, that is, the x_6-equation. Thus, if the x_6-equation had been underidentified when there was only one indicator of x_0^*, obtaining additional, "well-behaved," indicators would enable us to identify that equation. Note that this equation becomes identified when there are enough indicators, regardless of whether there are correlations among the errors in the indicators. (Recall that our use of x_1 and x_2 as instrumental variables for the x_6-equation did not depend on whether $\sigma_{de} = 0$.) However, when we come to an equation in which the unobserved variable appears as an explanatory variable (a direct cause of the endogenous variable), the situation is different. As we saw, our list of instruments for the x_5-equation depends on whether or not $\sigma_{de} = 0$. More generally, if the errors in the indicators are uncorrelated, we gain instrumental variables by adding (well-behaved) indicators. But if they are all intercorrelated, there is no such gain. Having more indicators will not, in this case, lead to the identification of a hitherto underidentified equation.

The assumption that errors in indicators are uncorrelated may, therefore, be quite useful. But it is also a very strong assumption. With only two indicators there is no way to test it. When there are three (all "well-behaved"), the only possibility of rejecting the assumption of uncorrelated errors is that the correlations among the three indicators will fall into the pattern of the "Heywood case" (Harman, 1967, pages 117–118). With four or more indicators, the hypothesis that all the indicators are "well-behaved" (in that they have no common causes other than the single unobserved variable x_0^*) and that their errors are uncorrelated is tantamount to the hypothesis that their intercorrelations are due to a single common factor. (See Harman, 1967, on factor analysis in general and Duncan, 1972, for a few simple cases.) As a matter of experience sociologists seldom have large sets of indicators with these attractive properties. As an empirical rule (exceptions granted), therefore, it is unlikely that an underidentified equation wherein an unobserved variable is one of the explicit causes will be rendered identifiable by finding several indicators of that variable.

Issues of identification aside, we have also seen that use of multiple indicators gives rise to complications in estimation. These may seem relatively benign, however, if one is able to take advantage of the new methods of estimation specifically contrived to deal with them (Jöreskog, 1970; Werts, Jöreskog, and Linn, 1973). But introduction of the additional indicator(s) does use resources that might be put into alternative use (for example, increasing sample size, or strengthening the model on the substantive side). In this connection, it should be remembered that with only a single indicator, even a highly fallible one (a high value of $\hat{\sigma}_{dd}/\hat{\sigma}_{11}$), we still do not have to suppose that its fallibility is imparting a bias to our estimates of structural coefficients. Only the precision, not the "validity," of those estimates is improved by lengthening our roster of indicators—provided, of course, the one indicator is "well behaved."

So much turns on the question of whether an indicator, or rather, its error structure, is "well behaved" that we should explore more systematically what this means. In essence, we are looking at the problem of "specification error" again, but this time with special reference to the problem of unobserved variables.

Let us complicate the illustrative model we have been working with so as to exemplify the case of indicators that are not "well behaved":

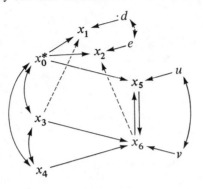

We will consider models including either the coefficient b_{13} or b_{26} or both, as will be made explicit by the x_1- and x_2-equations. The x_5 and x_6 equations remain as they were.

We begin with

$$x_1 = x_0^* + b_{13}x_3 + d$$

$$x_2 = ax_0^* + e$$

(blanking out the arrow from x_6 to x_2). Now, x_1 is no longer a "clean" (if fallible) measure of x_0^*. Instead it is "contaminated" by another exogenous variable in the model. We see why "validity of indicators" must be investigated with reference to a particular model. If x_3 were not in the model or were uncorrelated with x_0^*, this form of contamination would not matter a great deal. We find, perhaps surprisingly, that it is not necessary to alter the previously stated specifications on the errors and disturbances. Indeed, most of our results carry over from the previous version of the model. The x_6-equation is overidentified, with $x_1, \ldots, x_4$ eligible as instruments. The x_5-equation is also overidentified, in the somewhat different sense that there is more than one way to "solve out" x_0^* in that equation, although the two exogenous variables, x_3 and x_4, are still available as instruments. Upon solving out x_0^* from the x_1- and x_2-equations, we have

$$x_2 = ax_1 - ab_{13}x_3 + e'$$

where

$$e' = e - ad$$

or

$$x_1 = \frac{1}{a}x_2 + b_{13}x_3 + d'$$

where

$$d' = d - \frac{e}{a}$$

and thus encounter the indeterminacy in two-stage estimation again, owing to the sensitivity of this method to the normalization rule. As before, optimal methods of estimation are beyond our scope, although it is clear that four instrumental variables $(x_3, \ldots, x_6)$ are available. The overidentifying restrictions are made more explicit by the following derivations from the x_1- and x_2-equations:

$$\sigma_{13} = \sigma_{03}^* + b_{13}\sigma_{33} \qquad \sigma_{23} = a\sigma_{03}^*$$

$$\sigma_{14} = \sigma_{04}^* + b_{13}\sigma_{34} \qquad \sigma_{24} = a\sigma_{04}^*$$

$$\sigma_{15} = \sigma_{05}^* + b_{13}\sigma_{35} \qquad \sigma_{25} = a\sigma_{05}^*$$

$$\sigma_{16} = \sigma_{06}^* + b_{13}\sigma_{36} \qquad \sigma_{26} = a\sigma_{06}^*$$

The simple proportionality of the earlier model no longer holds. Instead, we have

$$\frac{\sigma_{13}}{\sigma_{23}} = \frac{1}{a} + \frac{b_{13}\sigma_{33}}{\sigma_{23}}$$

$$\frac{\sigma_{14}}{\sigma_{24}} = \frac{1}{a} + \frac{b_{13}\sigma_{34}}{\sigma_{24}}$$

$$\frac{\sigma_{15}}{\sigma_{25}} = \frac{1}{a} + \frac{b_{13}\sigma_{35}}{\sigma_{25}}$$

$$\frac{\sigma_{16}}{\sigma_{26}} = \frac{1}{a} + \frac{b_{13}\sigma_{36}}{\sigma_{26}}$$

putting the overidentifying restrictions in the form of an (exact) linear relation of the left-hand ratio of covariances to the right-hand ratio. This makes informal inspection of sample evidence bearing on the validity of the overidentifying restrictions easy to carry out. But a formal test is beyond our scope.

What one might find, therefore, is that this model offers an improved specification of the mechanism giving rise to the indicators—supposing, that is, that the first model was found wanting on that score while the present one is acceptable. The estimates for the x_5-equation would presumably be improved as a consequence. But two-stage estimates of the x_6-equation would not be affected. (Other methods—the so-called "system" methods, as distinct from a "single-equation" method like 2SLS or the two-stage method mentioned in this chapter—would, however, be favorably affected by the corrected specification of the model.)

We consider next the version of our illustrative model in which the contamination of an indicator arises from an endogenous rather than an exogenous variable (ignore the dashed arrow from x_3 to x_1 while retaining the one from x_6 to x_2):

$$x_1 = x_0^* + d$$
$$x_2 = ax_0^* + b_{26}x_6 + e$$

"Solving out" x_0^*,

$$x_2 = ax_1 + b_{26}x_6 + e'' \quad \text{where} \quad e'' = e - ad$$

We still have four variables $(x_3, \ldots, x_6)$ eligible as instruments. But for the reason noted before, a more complex method of estimation than two-stage regression will be desirable. Overidentifying restrictions are implied by

$$\sigma_{13} = \sigma_{03}^* \qquad \sigma_{23} = a\sigma_{03}^* + b_{26}\sigma_{36}$$
$$\sigma_{14} = \sigma_{04}^* \qquad \sigma_{24} = a\sigma_{04}^* + b_{26}\sigma_{46}$$
$$\sigma_{15} = \sigma_{05}^* \qquad \sigma_{25} = a\sigma_{05}^* + b_{26}\sigma_{56}$$
$$\sigma_{16} = \sigma_{06}^* \qquad \sigma_{26} = a\sigma_{06}^* + b_{26}\sigma_{66}$$

Again, we see that there must be an exact linear relationship between certain ratios of covariances:

$$\frac{\sigma_{23}}{\sigma_{13}} = a + \frac{b_{26}\sigma_{36}}{\sigma_{13}}$$

$$\frac{\sigma_{24}}{\sigma_{14}} = a + \frac{b_{26}\sigma_{46}}{\sigma_{14}}$$

$$\frac{\sigma_{25}}{\sigma_{15}} = a + \frac{b_{26}\sigma_{56}}{\sigma_{15}}$$

$$\frac{\sigma_{26}}{\sigma_{16}} = a + \frac{b_{26}\sigma_{66}}{\sigma_{16}}$$

Informal testing of the overidentifying restrictions is straightforward, but a formal test is beyond our scope.

The changed pattern of contamination has not affected the identification of the x_5-equation; x_3 and x_4 remain eligible as instruments. But we shall still have to resolve the implicit issue of weighting encountered when "solving out" x_0^* from the x_5-equation.

In regard to the x_6-equation, we no longer have $E(x_2 v) = 0$. Instead, as we find by multiplying through the new x_2-equation by v,

$$E(x_2 v) = aE(x_0^* v) + b_{26} E(x_6 v) + E(ev)$$

$$= b_{26}\sigma_{6v}$$

(since the other two covariances are still zero). Therefore, x_2 is not available as an instrument. But this is not fatal, since the x_6-equation was originally overidentified. Loss of one instrumental variable makes it just identified. We may estimate its coefficients by the IV method, with x_1, x_3, and x_4 as instruments.

We have seen that distinct consequences arise, according to whether the source of contamination is an exogenous or an endogenous variable. The strategy for exploring these consequences should now be familiar enough for you to attack the following

Exercise. *Consider a model represented by the previous diagram, but now suppose that both b_{13} and b_{26} are present. Investigate identifiability of the x_5- and x_6-equations and the parameters that govern the behavior*

of the indicators x_1 and x_2. Point out features of the model that give rise to problems in estimation that cannot be solved by the two-stage procedure suggested earlier and thus go beyond the scope of this book.

The examples and exercise cover only a few special cases. The reader should educate his intuition concerning problems of this kind by designing models to exemplify other cases. For example, how are the identification and estimation of the x_6-equation affected if x_5 is the source of contamination of one of the indicators? It seems possible to generalize to this extent. If a model is heavily overidentified, then quite a bit of quite "nasty" contamination of exogenous variables (only exogenous variables have been considered so far) is tolerable. However, every instance of contamination exacts a price in terms of the number of instrumental variables and number of overidentifying restrictions. The consequences, however, are not necessarily the same for the several equations of the model. Both the effect on identification and the implications for a strategy of estimation have to be studied carefully for each new model. Another caution: A situation in which the two (or more) indicators have the same pattern of contamination is tricky. Suppose, for example, that both x_1 and x_2 are functions of x_3 as well as x_0^*; our equations would read:

$$x_1 = x_0^* + b_{13}x_3 + d$$
$$x_2 = ax_0^* + b_{23}x_3 + e$$

"Solving out" x_0^* we have

$$x_2 = ax_1 - ab_{13}x_3 + b_{23}x_3 + e - ad$$
$$= ax_1 + (b_{23} - ab_{13})x_3 + e - ad$$

We can hope to estimate only one distinct coefficient for x_3 (even with the more advanced methods that are beyond the scope of this book!). Hence, we cannot distinguish between b_{23} and ab_{13}.

Exercise. *Determine whether this gives rise to unmanageable problems in connection with estimating the x_5- and x_6-equations.*

To illustrate complications that arise from multiple indicators of endogenous variables, we modify the model in Chapter 5 in such a way that each endogenous variable is unobserved but has two indicators:

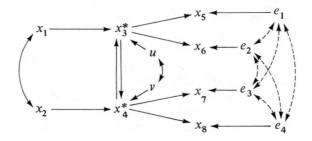

Our theory of how the indicators behave is expressed in four equations

$$x_5 = x_3^* + e_1$$
$$x_6 = a_1 x_3^* + e_2$$
$$x_7 = x_4^* + e_3$$
$$x_8 = a_2 x_4^* + e_4$$

We leave open temporarily the question of whether the errors should be assumed to be uncorrelated. The substance of the model is given by the two nonrecursive equations:

$$x_3^* = b_{31} x_1 + b_{34} x_4^* + u$$
$$x_4^* = b_{42} x_2 + b_{43} x_3^* + v$$

Exogenous variables are specified to have zero covariances with disturbances and measurement errors. Also, the unobserved variables have zero covariances with measurement errors. Hence (verify this) $E(e_h u) = E(e_h v) = 0$, $h = 1, \ldots, 4$; but $E(uv) \neq 0$.

We can easily "solve out" x_3^* from the x_5- and x_6-equations (the parallel discussion for x_4^* in the x_7- and x_8-equations can be supplied by relabelling subscripts):

$$x_6 = a_1 x_5 + e_2 - a_1 e_1$$

Two instrumental variables are available (x_1 and x_2), so that a_1 is overidentified. But in view of the indeterminacy of the normalization, two-stage regression has drawbacks as a method of estimation. Making explicit the overidentifying restrictions:

$$\sigma_{15} = \sigma_{13}^* \qquad \sigma_{16} = a_1 \sigma_{13}^*$$

$$\sigma_{25} = \sigma_{23}^* \qquad \sigma_{26} = a_1 \sigma_{23}^*$$

$$\sigma_{17} = \sigma_{14}^* \qquad \sigma_{18} = a_2 \sigma_{14}^*$$

$$\sigma_{27} = \sigma_{24}^* \qquad \sigma_{28} = a_2 \sigma_{24}^*$$

Hence,

$$a_1 = \frac{\sigma_{16}}{\sigma_{15}} = \frac{\sigma_{26}}{\sigma_{25}}$$

$$a_2 = \frac{\sigma_{18}}{\sigma_{17}} = \frac{\sigma_{28}}{\sigma_{27}}$$

A simple informal test is obvious, but a formal test is beyond our scope.

At this point, let us consider the implications of the specification of uncorrelated measurement errors, $E(e_h e_j) = 0$; $h, j = 1, \ldots, 4$; $h \neq j$. We multiply each of the four indicator equations by each of the others, to obtain:

$$\sigma_{56} = a_1 \sigma_{33}^* + \sigma_{e_1 e_2}$$

$$\sigma_{78} = a_2 \sigma_{44}^* + \sigma_{e_3 e_4}$$

$$\sigma_{57} = \sigma_{34}^* + \sigma_{e_1 e_3}$$

$$\sigma_{58} = a_2 \sigma_{34}^* + \sigma_{e_1 e_4}$$

$$\sigma_{67} = a_1 \sigma_{34}^* + \sigma_{e_2 e_3}$$

$$\sigma_{68} = a_1 a_2 \sigma_{34}^* + \sigma_{e_2 e_4}$$

Setting all the error covariances to zero, there are five distinct parameters on the right-hand side and six covariances of observable variances on the left-hand side. The parameters are clearly overidentified. What we have in our four equations is a particular factor analysis model, methods of estimation for which would have to be sought in the literature of that subject. We see, however, that estimates of a_1 and a_2

obtained by this route will not be the same as those obtained by using instrumental variables. That is, still another kind of overidentifying restriction is implied by the availability of these two distinct solutions. At this point, an investigator might well ask herself whether she needs so strong a theory about her indicators. Relinquishing the assumption of uncorrelated measurement errors will cost her the possibility of estimating the variances and covariance of the unobserved variables. But there is no clear need for these estimates in any case, since the structural coefficients in the x_3^*- and x_4^*-equations can be estimated without knowing them. Henceforth, we shall assume that any pair of e's may have a nonzero covariance.

Turning to our structural equations, we see that the estimation problem is complicated by the fact that both x_3^* and x_4^* appear in both equations and that there are alternative ways to "solve out" the unobserved variables. On the other hand, it is clear that once this is done—we will assume that the appropriate "weighting" of the x_5-vis-à-vis the x_6-equation and the x_7- vis-à-vis the x_8-equation will, in effect, be determined in the process of estimating a_1 and a_2—each equation is identified, since there are two structural coefficients in each and two instrumental variables are available. As is the case throughout this chapter, we consider that the reader has fair warning if we state that a method of estimation substantially more complicated than two-stage regression will be required for this problem.

The advantages and disadvantages of multiple indicators for endogenous variables turn out to be much the same as those noted for exogenous variables earlier in the chapter. On the one hand, they do not necessarily help with the identification problem, depending on what one can assume about correlations among errors in the indicators. A model that is underidentified with one indicator per endogenous variable may remain so with many indicators. Estimation is more complicated and costly than with a single indicator. On the other hand, one does secure the chance to make a partial test of whether the indicators are "well behaved." If it turns out that they are, estimates of structural coefficients are more precise (ordinarily they will have smaller sampling errors) than they would be with one indicator per variable. But there is no other sense in which those estimates are improved.

However fascinating may be the problems that arise with multiple indicators, we have to recognize that sometimes they merely complicate if not obscure what is surely the more fundamental problem: proper specification of our models in substantive terms.

FURTHER READING

In this chapter it has been necessary to mention repeatedly "methods that go beyond the scope of this book." The most important references are two papers of Jöreskog (1970, and his contribution to Goldberger and Duncan, 1973, Chapter 5). These are highly technical articles. A brief and more accessible exposition, with several examples, is given by Werts, Jöreskog, and Linn (1973). For a critique of earlier sociological writing on the multiple indicators problem and an explication of the meaning of "efficient estimation" in this context, see Hauser and Goldberger (1971).

11

Form and Substance of (Sociological) Models

Like most subjects, the study of structural equation models can be divided into two parts: the easy part and the hard part. The distinction is roughly equivalent to the difference between "talking about it" and "doing it." You will have observed long since that this book is only an introduction to the easy part. (Actually, even that part gets difficult enough—if this is not a contradiction in terms—unless you attack it with a firm grasp on the essential tools, like matrix algebra and mathematical statistics, which are not assumed to be available to the reader of this book.)

Few books are written about the "hard part" of our subject, attempting to teach you how to "do it." A logical paradox, rather than the reticence of authors, is the probable explanation of this gap in the literature. Ultimately, for the author to tell you how to "do it" in sufficient detail that you could henceforth do it yourself with no further mental exertion would require him actually to "do it" himself ahead of you. And, you can be sure, if he knew how to do that, he would have long since done it! There is no formula for "doing" science. If such a formula were, paradoxically, to be discovered, we could program it for a computer and have done with the tedious business of thinking for ourselves.

Of course, one can hope to gain some insights or inspiration from studying the successful or unsuccessful results of those who have tried to "do it." Sociological examples of structural equation models have appeared with some frequency in the periodical literature since about 1966 and in such collections or symposia as Blalock (1971) and Goldberger and Duncan (1973). The student will do well to assume that many of these examples are defective in one way or another. (Some useful warnings are posted in Goldberger, 1972b.) Some of them actually contain elementary formal errors. All are debatable as substantive contributions to social science (as is, of course, any empirical investigation), and the more suggestive and stimulating papers may be the more debatable ones. But even if one can describe, after the fact, what it is that makes a good example good (or a bad example bad, even if instructive with regard to what not to do), one does not have a recipe for creating a new model. There is a certain amount of "wisdom" to be attained by close study of previous work, along with possible insight or inspiration, but it typically takes a negative form: an ability to sense dead ends or traps to be avoided. Any effort to encapsulate such wisdom leads to the vacuous advice that you shouldn't repeat the mistakes of your predecessors.

At the risk of vacuity, let me offer one piece of "wisdom." Do not undertake the study of structural equation models (or, for that matter, any other topic in sociological methods) in the hope of acquiring a technique that can be applied mechanically to a set of numerical data with the expectation that the result will automatically be "research." Over and over again, sociologists have seized upon the latest innovation in statistical method, rushed to their calculators or computers to apply it, and naively exhibited the results as if they were contributions to scientific knowledge. The lust for "instant sociology," the superstition that it is to be achieved merely by a complication if not perfection of formal or statistical methods, and the instinct to suppose that any old set of data, tortured according to the prescribed ritual, will yield up interesting scientific discoveries—all these pathological habits of thought are grounded (if at all) in the fallacy of induction. (If you do not know what the fallacy of induction is, you have some reading to do in the philosophy of science.)

It might appear that the literature of sociological investigations using structural equation models provides just another episode of seizing upon the latest methodological fad in the incessant quest for the formula for instant sociology. Your present author would not care to defend too vigorously the contrary view. But if any of these sociological examples *are* contributions to science (and not merely exercises in quantitative technique), it is because the models rest on creative, substantial, and sound sociological theory. (Needless to say, there is a division of labor in inquiry, so that the originators of significant sociological ideas need not be the same investigators as those who succeed in embodying those ideas in models.) A benign interpretation of the flourishing of structural equation models in recent sociology would be, therefore, that these models provided a tool that only facilitated the genuinely scientific inquiry that sociologists were already engaged in and, indeed, that the models responded to an implicit need for formalisms that would help in maintaining order and coherence in increasingly complicated lines of investigation and theorizing.

In short, your models can be no better than your ideas, even if trying to put your ideas in the form of models *may* (if yours is the kind of mind that works this way, but not otherwise) stimulate the flow of ideas.

What we have to say about form and substance of models, therefore, is not really a set of "how to" instructions. Rather, we offer a few reflections about the logic of models like those described in this book. Or, if you prefer, these are remarks addressed to the semantics (rather than, as up to this point, the syntax) of the language in which we talk about these models, especially such key terms as variables and coefficients, units and populations, structural form and disturbances.

It does not seem to be possible to give a definition of "structural form" that is other than circular. The structural form of the model is that parameterization—among the various possible ones—in which the coefficients are (relatively) unmixed, invariant, and autonomous. How do you know if you have written a model in its structural form, rather than in some other form? Well, if the coefficients in the model are indeed relatively invariant across populations, somewhat autonomous, and not inseparable mixtures of the coefficients that "really" govern how the world works—then your model is actually in its "structural" form. Thus, we cannot tell just by studying the mathematical properties of a model whether it is a "structural equation model" or not, although we can determine whether it is recursive or nonrecursive, identified or underidentified, and so on. If an investigator submits a model and asserts that it is a structural equation model, then we may take it that he has reasons to believe that the coefficients are unmixed, invariant, and autonomous; and we should inquire what those reasons are and assess (in the light of the best available theory on the subject) their plausibility.

A strong possibility in any area of research at a given time is that there are *no* structural relations among the variables currently recognized and measured in that area. Hence, whatever its mathematical properties, no model describing covariation of those variables will be a structural model. What is needed under the circumstances is a theory that invents the proper variables. (There were no structural equation models—or their functional equivalent—in genetics until Mendel's theory was available, although there were plenty of "studies" of parent–offspring resemblance. There were no structural equation models for the epidemiology of malaria until the true agent and vector of the disease were identified, although there were plenty of correlations between prevalence of the disease and environmental conditions.)

How should we state these formidable requirements, that *structural* coefficients be unmixed, invariant, and autonomous? At least the first of these notions is easily illustrated using our familiar "specification error" rhetoric. Suppose the true structural model is

$$x_3 = b_{31}x_1 + b_{34}x_4 + u$$
$$x_4 = b_{42}x_2 + b_{43}x_3 + v$$

where the b's are the coefficients that actually govern how the world works, that is, the structural coefficients, while u and v are disturbances with the usual specification. Then, the following is also a valid model:

$$x_3 = a_{31}x_1 + a_{32}x_2 + u'$$
$$x_4 = a_{41}x_1 + a_{42}x_2 + v'$$

where

$$a_{31} = \frac{b_{31}}{\Delta}$$

$$a_{32} = \frac{b_{34}b_{42}}{\Delta}$$

$$a_{41} = \frac{b_{43} b_{31}}{\Delta}$$

$$a_{42} = \frac{b_{42}}{\Delta}$$

$$u' = \frac{u + b_{34} v}{\Delta}$$

$$v' = \frac{b_{43} u + v}{\Delta}$$

$$\Delta = 1 - b_{34} b_{43}$$

But note that each of the a's is a *mixture* of three or four b's. Suppose, for example, that one of the b's changes as we go from one population to another while the others do not. We incur no logical contradiction in making this assumption. But if this happens, at least two of the a's will change. In this sense, the b's are more *autonomous* than the a's. We actually do incur a contradiction by supposing that one of the a's can change while the others do not, because this would mean that several b's had changed and, thereby, some other a's. Calling the model, in the version written in terms of the a's, a "structural" model is a "specification error," albeit a different kind of error from those usually so labelled. Yet, this model is a perfectly valid description of what is going on in the population in which it applies.

Let us turn the premise of the discussion around. Suppose, now, that the model written in terms of the a's is actually a (recursive) structural model. In that event, it is the b's that are "mixtures" of structural coefficients:

$$b_{31} = \frac{a_{31} a_{42} - a_{32} a_{41}}{a_{42}} \qquad u = u' - \frac{a_{32}}{a_{42}} v'$$

$$b_{42} = \frac{a_{31} a_{42} - a_{32} a_{41}}{a_{31}} \qquad v = v' - \frac{a_{41}}{a_{31}} u'$$

$$b_{34} = \frac{a_{32}}{a_{42}}$$

$$b_{43} = \frac{a_{41}}{a_{31}}$$

Is it then the nonrecursive model, written in terms of the b's, or the recursive model, written in terms of the a's, that is the "real" structural model? The answer is that invariance is where you find it—in the world or in your theory—and no analysis of a model's formal properties can determine whether its coefficients are invariant.

It might be thought that an easy answer, in some cases, would emerge from counting parameters. Other things being equal, we are often told, science follows the rule of parsimony. It prefers the formulation with fewer to the one with more parameters. Indeed, we have seen (Chapter 7) that in the case of an overidentified nonrecursive model, there are more reduced-form than structural coefficients, so that the structural form would be preferred on grounds of economy of description. Yet this could hardly be a decisive criterion, for in that case, the problem of *under*identification would never arise. One would simply have recourse to the reduced form, since its parameters are not only identified but are fewer in number than those of the structural form. We get into the identification problem (Chapter 6) precisely because we want to express the model in terms of what we think are *structural* coefficients. If it then turns out that the model is underidentified, we may not see how to implement it in an empirical context. But at least we do not deceive ourselves by taking the easy way out—doing statistics (merely calculating regressions for the reduced form) rather than science (broadening or elaborating our theory, or improving the design of our research, so as to achieve an identified model).

An especially disconcerting possibility is that neither of the formulations above—in terms of a's or b's—will provide a model that is really "structural," in the strong sense of that term being used in this chapter. Suppose this is the true model:

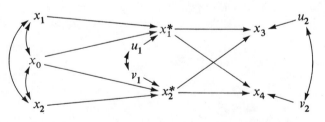

Perhaps x_1^* and x_2^* are latent variables, quite "real" in terms of our theory but (in the present state of knowledge) not observable except by way of their consequences, x_3 and x_4. The equations corresponding to the diagram are

$$x_1^* = c_{10}x_0 + c_{11}x_1 + u_1$$

$$x_2^* = c_{20}x_0 + c_{22}x_2 + v_1$$

$$x_3 = c_{31}x_1^* + c_{32}x_2^* + u_2$$

$$x_4 = c_{41}x_1^* + c_{42}x_2^* + v_2$$

To do anything with the model statistically, we must "solve out" the unobserved variables from the last two equations. If we simply substitute the first two equations into the last two, we find

$$x_3 = (c_{31}c_{10} + c_{32}c_{20})x_0 + c_{31}c_{11}x_1 + c_{32}c_{22}x_2 + u_2$$
$$+ c_{31}u_1 + c_{32}v_1$$

$$x_4 = (c_{41}c_{10} + c_{42}c_{20})x_0 + c_{41}c_{11}x_1 + c_{42}c_{22}x_2 + v_2$$
$$+ c_{41}u_1 + c_{42}v_1$$

This is the reduced-form model, with

$$a_{30} = c_{31}c_{10} + c_{32}c_{20}$$

$$a_{31} = c_{31}c_{11}$$

$$a_{32} = c_{32}c_{22}$$

$$a_{40} = c_{41}c_{10} + c_{42}c_{20}$$

$$a_{41} = c_{41}c_{11}$$

$$a_{42} = c_{42}c_{22}$$

$$u' = u_2 + c_{31}u_1 + c_{32}v_1$$

$$v' = v_2 + c_{41}u_1 + c_{42}v_1$$

Note that the a's are all "mixtures" of the c's—the real "structural" coefficients in this example, which are not identified, although certain ratios of coefficients, c_{41}/c_{31} and c_{32}/c_{42}, are identified.

An alternative manipulation is to solve for the unobserved variables in the second pair of equations in terms of the endogenous variables.

This yields

$$x_1^* = \frac{c_{42}x_3 - c_{32}x_4 - (c_{42}u_2 - c_{32}v_2)}{c_{31}c_{42} - c_{32}c_{41}}$$

$$x_2^* = \frac{c_{31}x_4 - c_{41}x_3 - (c_{31}v_2 - c_{41}u_2)}{c_{.1}c_{42} - c_{32}c_{41}}$$

We now equate the above expression for x_1^* to $c_{10}x_0 + c_{11}x_1 + u_1$ (the first equation in the model) and the expression for x_2^* to $c_{20}x_0 + c_{22}x_2 + v_1$ (from the second equation). After rearranging terms we have (with $D = c_{31}c_{42} - c_{32}c_{41}$)

$$x_3 = \frac{Dc_{10}}{c_{42}}x_0 + \frac{Dc_{11}}{c_{42}}x_1 + \frac{c_{32}}{c_{42}}x_4 + u_2 + \frac{D}{c_{42}}u_1 - \frac{c_{32}}{c_{42}}v_2$$

$$x_4 = \frac{Dc_{20}}{c_{31}}x_0 + \frac{Dc_{22}}{c_{31}}x_2 + \frac{c_{41}}{c_{31}}x_3 + v_2 + \frac{D}{c_{31}}v_1 - \frac{c_{41}}{c_{31}}u_2$$

which is seen to be a just-identified nonrecursive model, once we adopt the notation,

$$b_{30} = \frac{Dc_{10}}{c_{42}}$$

$$b_{31} = \frac{Dc_{11}}{c_{42}}$$

$$b_{34} = \frac{c_{32}}{c_{42}}$$

$$b_{40} = \frac{Dc_{20}}{c_{31}}$$

$$b_{42} = \frac{Dc_{22}}{c_{31}}$$

$$b_{43} = \frac{c_{41}}{c_{31}}$$

$$u = u_2 + \frac{Du_1}{c_{42}} - \frac{c_{32}v_2}{c_{42}}$$

$$v = v_2 + \frac{Dv_1}{c_{31}} - \frac{c_{41}u_2}{c_{31}}$$

Like the a's, the b's are mixtures of coefficients in the original model. The comparison between the two derived models in purely formal terms can hardly tell us whether the a's or the b's are more likely to be invariant. Perhaps the theory underlying the original model could be of aid in this regard.

[A side benefit of this example is that it sheds some light on a question that has mystified some sociologists—why do we often encounter negative estimates of the covariance between disturbances in nonrecursive models? In the example, we make the standard assumption that disturbances are uncorrelated with predetermined variables, whence $E(u_1 u_2) = E(u_1 v_2) = E(v_1 u_2) = E(v_1 u_2) = 0$, although $E(u_1 v_1) \neq 0$ and $E(u_2 v_2) \neq 0$. Evaluating $E(uv)$ in terms of the definition of u and v, we find

$$\sigma_{uv} = \left(1 + \frac{c_{32} c_{41}}{c_{31} c_{42}}\right) E(u_2 v_2) + \frac{D^2}{c_{31} c_{42}} E(u_1 v_1) - \frac{c_{41}}{c_{31}} E(u_2^2) - \frac{c_{32}}{c_{42}} E(v_2^2)$$

There is no particular reason why negative could not outweigh positive terms in this expression. It is also worth noting that σ_{uv} is not necessarily zero, even if $E(u_1 v_1) = E(u_2 v_2) = 0$, that is, even if the original model is fully recursive. This may help to explain why the specification of uncorrelated disturbances rarely seems appropriate for a nonrecursive model, especially if the latter is assumed to arise from an unobservable recursive process.]

Let us play one more round of this little game. Consider again the reduced-form version of our model, which (as we saw on page 155) can be written:

$$x_3 = a_{30} x_0 + a_{31} x_1 + a_{32} x_2 + u'$$

$$x_4 = a_{40} x_0 + a_{41} x_1 + a_{42} x_2 + v'$$

This is not a fully recursive model, for the reason that $E(u'v') \neq 0$. Moreover, neither $E(x_3 v')$ nor $E(x_4 u')$ is zero. We multiply the x_3-equation through by a constant K, subtract the result from the x_4-equation, and rearrange the terms, to obtain

$$x_4 = (a_{40} - Ka_{30})x_0 + (a_{41} - Ka_{31})x_1 + (a_{42} - Ka_{32})x_2 + Kx_3$$
$$+ v' - Ku'$$

Using a new notation, in which

$$d_{30} = a_{30}$$
$$d_{31} = a_{31}$$
$$d_{32} = a_{32}$$
$$d_{40} = a_{40} - Ka_{30}$$
$$d_{41} = a_{41} - Ka_{31}$$
$$d_{42} = a_{42} - Ka_{32}$$
$$d_{43} = K$$
$$u'' = u'$$
$$v'' = v' - Ku'$$

we may summarize our result as a system of two equations:

$$x_3 = d_{30}x_0 + d_{31}x_1 + d_{32}x_2 + u''$$
$$x_4 = d_{40}x_0 + d_{41}x_1 + d_{42}x_2 + d_{43}x_3 + v''$$

Moreover, if we take care to choose the value of K so that

$$K = \frac{E(u'v')}{E[(u')^2]}$$

this will be a fully recursive system; that is, $E(x_h u'') = 0$, $h = 0, 1, 2$; $E(x_h v'') = 0$, $h = 0, \ldots, 3$; and $E(u''v'') = 0$.

Exercise. *Show that these properties hold as a consequence of the value of K.*

From a strictly mathematical point of view, there is simply no way to choose between the reduced form (written in terms of a's), the non-recursive or simultaneous form (written in terms of b's), and the fully recursive form (written in terms of d's). Each of them, unfortunately, "mixes up" the coefficients of the "true" model (the c's).

We must steel our resolve not to counsel with statisticians or mathematicians when trying to decide which (if any) of the mathematically interchangeable ways of stating our model is scientifically relevant.

That is not their job, although the scientist does appreciate their help in showing him how various parameterizations are always possible if only one is knowledgeable about the properties of mathematical transformations. As to how one *does* confirm or disconfirm—presumably the latter is the realistic possibility—the assumption that coefficients are invariant, we can only state that there are no special techniques associated with structural equation models. The task of challenging a theory with any possibly relevant evidence calls for as much imagination and ingenuity here as in any kind of empirical testing of theory. In short, invariance and autonomy are not formal properties of a model, but substantive assumptions we make in motivating our work with a model.

We have already hinted that the basic difficulty with a model may reside, not in its mathematical form, but in the very variables that enter into it. Sociologists have worried a good deal about this problem. The interest in multiple indicators (noted in Chapter 10) is one aspect of the concern with definition of variables. There has also been a lively interest in whether structural equation models are suited to the use of variables whose level of measurement is "nominal" or "ordinal," rather than "interval" or "ratio." The false issue in much of the discussion on this matter is whether these different types of scale are suited to one or another kind of *statistical* manipulation. But the real question is (should be) whether, having done the statistics, one has a *scientifically* useful result.

It is true that models of the kind presented in this book treat *dependent* variables as being measured on an interval (if not ratio) scale. If one must perforce consider a variable for which only ordinal measurement can be claimed, what damage is done in assigning numbers to the various grades of that scale and henceforth manipulating those numbers as if they arose from measurements on an interval scale? In college, for example, instructors grade students on the ordinal scale, A, B, C, D, F, and the registrars assign to these grades the numbers 4, 3, 2, 1, 0, respectively, in order to compute the "grade-point average." Clearly, such assignments are arbitrary. One might equally well use the numbers 16, 9, 4, 1, 0 in computing grade-point averages—unless, through convention or habituation, students and faculty come to feel that the difference between an A and a B is equal to the difference between a C and a D, and so on.

In introducing an arbitrarily scored ordinal variable into a model, we do not stop with computing its average value. Instead, that variable enters into a number of linear equations (unless some other functional form is specified). Hence, in selecting numbers to assign to the grades, we should choose them in such a way (if possible) that the scores do bear a linear relationship to variables causing and/or affected by the ordinal variable. Assuming success in this venture, we do not have to claim that the scores are the true scale of measurement, but only that (in effect) they result from some monotonic transformation of the true scale. Since the introduction of such transformations is an accepted scientific procedure for linearizing relationships, even if the variables are originally measured on an interval scale, there is no obvious reason why one should feel nervous in working with a variable that can be assumed to be the outcome of such a transformation. Indeed, one perfectly good meaning of "interval" in the first place is that equal changes on a causal variable in different parts of its range produce equal changes in the effect variable. One consequence of this point of view is that the issue with regard to measurement is not resolved before a variable is entered into the model, but in the very process of entering it. Those scales that "work," in the sense of providing measurements which are well behaved—the notion of invariance again—in significant models, are those that we come to accept as natural.

There is, however, another aspect to the problem. Many "ordinal" scales in sociology provide for only very coarse measurement and limit the possible range of variation to as few as two, three, four, or five grades or steps on the scale. This tends, on the one hand, to introduce measurement error—and not necessarily error that is well behaved in the sense assumed throughout Chapter 9—because of the imprecision in applying the scale to real-world phenomena. Even more important, it may invalidate the crucial assumption of all our models that the disturbance is distributed independently of the exogenous variable(s).

This is especially clear if the ordinal scale is merely a dichotomy. We may suppose, without losing generality, that its two possible values are zero and unity. Now, let us reconsider the model of Chapter 1,

$$Y - \overline{Y} = bx + u$$

on the assumption that Y is dichotomous, while x is, for all practical purposes, continuous. Possible values for Y are 0 and 1. Possible values for $\hat{Y} = \overline{Y} + bx$ are not limited in this way. They may fall outside the range 0 to 1 (an intrinsically absurd result) and can take on any value within it. Suppose that for $x = K_0$, $\hat{Y} = .5$; then $\sigma_{uu} = (.5) \times (.5) = .25$. But if, for $x = K_0 + K_1$, $\hat{Y} = .8$, and for $x = K_0 - K_1$, $\hat{Y} = .2$, then (in each case) $\sigma_{uu} = (.8)(.2) = .16$. Thus, the homoskedasticity assumption is violated, and in an intrinsic way. When the dependent variable is dichotomous, one of the basic specifications on the disturbance is self-contradictory. Presumably, something of the same kind of violation occurs whenever the dependent variable is severely limited in its range, by having only a few, though more than two, ordered categories.

We should note that this problem does not occur if the variable in question is exogenous. Indeed, in that event, there is no difficulty even if the categories are not ordered. The wellknown statistical transformation of a nominal variable into a set of dummy variables will work quite satisfactorily here, whether the model is recursive or nonrecursive. (For an example of a model with a set of exogeneous dummy variables see Duncan and Duncan, 1968.)

There seem to be two broad strategies for coping with the problem posed by variables that are measured with only a small number of categories. One is to improve the measurement procedure. In many cases, we work with a three- or four-point scale only for reasons of convenience. It may seem too expensive, for example, to ask several questions to elicit a reasonably precise expression of opinion, so we satisfy ourselves with a coarse categorization. In the past, sociologists working in the tradition of survey research have too often taken the easy way out on this issue. It is the policy of an ostrich to assume that we can ignore indefinitely the basic work on theory of measurement being done in other fields. Once we learn to take the job of measurement seriously, the problem under discussion here will tend to resolve itself. [Two approaches to measurement, among others, that should be better known to sociologists are exemplified in articles by Rasch (1968) and Stevens (1960).]

The other strategy presumes that, in some cases, qualitative categories do not arise from coarse scaling of a basically continuous variable, but are intrinsic to the attribute being observed. In that event, one does well to inquire whether our models, rather than our measurement procedures, need to be reconsidered. It so happens that there is a strong tradition of so-called survey analysis that works primarily with qualitative variables. There is a well-developed causal language for survey analysis, especially in the work of Lazarsfeld and associates (1972), who also provided us with prototypical models of latent structures for qualitative variables. Until very recently, however, the statistical implementation of survey analysis and latent structure models was not on a sound footing. But now we have, especially in the contributions of Goodman (1972; 1974), the statistical tools needed to produce the functional equivalent of structural equation models for qualitative variables. It should also be mentioned that there are various other approaches to model building that, broadly speaking, use a kind of causal reasoning that is quite congenial to the spirit of structural equation models, even though the mathematical and statistical techniques are different. These include sociodemographic models (Stone, 1971; Keyfitz, 1971), stochastic process models (Coleman, 1964; Boudon, 1967), and certain kinds of simulation models.

Whether the variables be quantitative or qualitative, the substantively problematic issue is the relationship among variables. In structural equation models, this relationship is expressed by the structural coefficient, a number whose dimensions are units of change in y per unit change in x. In a formal exposition of the properties of a model, all we need to assume is that there is such a number. In substantive applications, the problem is what that number means—what are "coefficients," anyway? It is important to recognize that coefficients with quite different kinds of theoretical meaning are all in the same mathematical form. Whether a purported coefficient even has an intelligible meaning is a substantive, not a formal problem.

A coefficient may be nothing more than a scale factor which converts a variable from one metric to another, like the (9/5) in the formula relating two temperature scales,

$$F° = 32° + (9/5)C°$$

If the formula is empirically derived, as in calibration experiments, there will be an error term as well as a coefficient and, possibly, a constant. (Of course, calibration may require a nonlinear formula with more than one coefficient.)

In models that purport to explain the behavior of individual persons, the coefficients could well take the form of units of response per unit of stimulus strength; the structural equation is, in effect, an S–R law. Suppose that the general form of this law is

$$R = a + bS + U$$

assuming that (if necessary) R and S have been transformed in such a way that the relationship is linear. We are here making the intercept explicit, that is, we drop the assumption made hitherto that the variables are expressed as deviations from their means.

We need to be explicit about the interpretation of this law. Using the subscript i to identify the individual, if we write,

$$R_i = a + bS_i + U_i$$

this means that every individual is subject to the same S–R law, $R = a + bS$; but when we subject an individual to a stimulus of magnitude S_i, his response, R_i, is subject to a random deviation, U_i, from the value, $\hat{R}_i$, implied by the law. This could be due to imprecision in the measurement of the response or to an intrinsic variability in behavior. The situation is like that depicted (for a very small sample of observations) in Fig. 11.1. On this graph R is on the Y-axis, S on the X-axis;

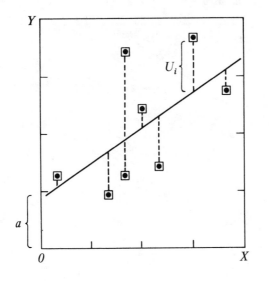

FIGURE 11.1

and the U_i are shown as deviations from the values predicted by the model for a given magnitude of X. The specification on the disturbance is that $E(u_i^2) = \sigma_{uu}$ for all i. This means, among other things, that in repeated independent experiments on the same individual, with S varying or fixed at a single value, the variance of the responses around the theoretical values would be σ_{uu}. Hence, in one experiment the individual might have a small value of U, in another a large value, and so on.

Another, quite different, interpretation of U_i is that it is a permanent characteristic of the ith individual. For any given magnitude of stimulus strength, some individuals will respond with relatively high values of R and others with relatively low values. But any individual, in independent experiments involving different magnitudes of S, will respond according to the linear S–R law with coefficient b. In this case, we could be more explicit by writing our model:

$$R_i = a_i + bS_i$$

This interpretation of the equation is depicted graphically in Fig. 11.2. The disturbances in the first interpretation have become intercepts in the second, to convey the notion that deviations from what must now be thought of as an *average* S–R law, $R = a + bS$, are real and per-

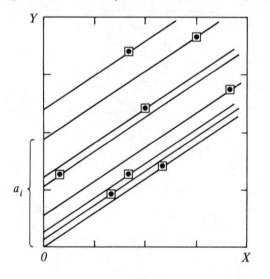

FIGURE 11.2.

manent individual differences and not merely transitory variations in behavior.

In some situations, it is impossible to distinguish empirically between these two situations. Note that the same data are used in Figs. 11.1 and 11.2. If we have one observation per individual, varying S from each individual to the next (whether systematically or randomly), then error in measuring R, intrinsic (intra-individual) variability in R, and individual differences (a_i) in level of R for fixed value of S are summed up in a single quantity, U_i. As long as we depend solely on inter-individual comparisons to estimate b, this will be the case. If the stimulus is administered, as it were, by nature on a once-in-a-lifetime basis (as is true of many sociological variables, regarded as stimuli), then our major strategy must be, in effect, to try to explain the variation in U_i. To the extent that we can accomplish this, by specifying traits or social characteristics of individuals which raise or lower R_i (apart from any intrinsic or adventitious correlation with S_i), we will be justified in thinking of U_i as at least partly reflecting individual differences.

The alternative strategy is to make multiple observations on the same individual causing or allowing S to vary between observations. If the observations are independent (there is, for example, no carry-over from one to another, due to learning or memory), this study design opens up the possibility of estimating a_i from data on the ith individual alone. It also opens up the possibilities of testing whether b is indeed constant from one individual to another and whether σ_{uu} is actually the same for all individuals. If the model is sustained against both these tests, the design permits one to make separate estimates of a_i and U_i. Indeed, a more explicit model is

$$R_{ij} = a_i + bS_{ij} + U_{ij}$$

where we are explaining the response of the ith individual upon the jth administration of the stimulus.

Many sociologists would probably disclaim interest in S–R laws or models featuring such relations. It might be a useful exercise for them to reflect on the question of what their coefficients do mean. One answer might be that the sociologist is interested in *societal* responses to individuals—or characteristics thereof—as stimuli. Then X_i is perhaps a measure of the individual's performance and Y_i his reward, with the (implicit or explicit) rule or convention relating the two being

$$Y_i = a + bX_i + U_i \qquad (a)$$

or

$$Y_i = a_i + bX_i \qquad (b)$$

Again, we are constrained to note that versions (a) and (b) are conceptually different though mathematically equivalent. Much of the controversy surrounding Jencks' *Inequality* (1972) turned on the author's conclusion that a major factor in the distribution of income is "luck," which fact he inferred from what he regarded as the relatively large magnitude of $\hat{\sigma}_{uu}/\hat{\sigma}_{yy}$. He had taken considerable pains to dispose of version (b) by including in his model a number of characteristics of individuals that a society might conceivably pay attention to, in allocating rewards; his actual model involved several X's, not just one. Without commenting on the merits of Jencks's conclusion we note that the "b" in his model—say, dollars of income annually per number of years of schooling completed—is hardly a coefficient that describes merely the behavior of the ith individual. After all, it is *someone else·* who pays him the income in recompense for his schooling—or, more plausibly, the economic performance that his schooling makes possible.

The "structures" that are described in the sociologist's "structural equation models" are, therefore, likely to be *social structures* in a nontrivial sense of the term. The coefficients may represent the way in which explicit social rules or regulations operate, or they may be unintended resultants of innumerable conventional actions. Individual differences—or unit differences, if the social units are other than persons—will be present in the model, whether explicitly in the a_i or (following the usual practice, as in earlier chapters of this book) in the U_i. We should expect their variance to be large. A complex social system can attend to or differentiate its response to a rather large number of individual characteristics, but not nearly so large as the number on which individuals actually differ. The same system will implicitly operate (at least in part) on a stochastic principle, selecting at random those individuals to be assigned this set of roles or that set of sanctions. The sociologist who despairs of his low R^2's would do well to ask himself if he would want it otherwise—would he care to live in the society so structured that his particular collection of variables accounts for 90% instead of 32% of the variance in Y?

The structures we hope to describe in our model are not fixed in the human gene pool nor dictated by the laws of physics. They can and do change. Much of the difficulty sociologists have had with the problem of social change stems from their indifferent success in sorting out which changes result from:

(*a*) alterations implicit in the very nature of the social structure and its initial conditions, as growth is implicit in the structure of the immature organism;

(*b*) variations due solely to differences in exogenous variables operating upon a constant structure (see discussion in Chapter 4, pages 56–57);

(*c*) fluctuations analogous to those of sampling from a given probability distribution; and

(*d*) structural changes as such, that is, change in the structural coefficients.

What we mean by "structural change" will, in practice, be relative to the state of the art in building models of social structures. If one is optimistic about future success in this venture, one would expect much that now passes for structural change to be reinterpreted as falling within another one of the above rubrics. In any case, as structural equation models (or the broader class of "causal" models of all kinds) become more numerous, we shall have to get much more specific about the "population to which the model applies," as it has been termed in this book.

We can conceive of two populations (or the same population at two points in time) in which the same structure holds—that is, the model is the same and the numerical values of the coefficients are the same. We can also imagine that the model is the same, but one or more coefficients are different. It is this latter kind of variation (over time, or between different social universes), in particular, that ought to be especially amenable to study with tools like those introduced in this book. If we can find situations or time periods where only a few coefficients differ, we may have some hope of coming to understand the impact of these differences—our models can help directly with that—and perhaps of making useful conjectures as to their origins.

FURTHER READING

Carlsson (1972) discusses "strategies of discovery and concealment" and suggests reasons why causal models in sociology are likely to be seriously misspecified.

References

Blalock, H. M. Jr. Four-variable causal models and partial correlations. *American Journal of Sociology*, 1962–1963, **68**, 182–194, 510–512 (Erratum). [Reprinted in Blalock (1971).]

Blalock, H. M. Jr. (Ed.) *Causal models in the social sciences*. Chicago: Aldine-Atherton, 1971.

Boudon, R. *L'analyse mathématique des faits sociaux*. Paris: Plon, 1967.

Cain, G. G., & Watts, H. W. Problems in making policy inferences from the Coleman Report. *American Sociological Review*, 1970, **35**, 228–242.

Carlsson, G. Lagged structures and cross-sectional methods. *Acta Sociologica* 1972, **15**, 323–341.

Christ, C. F. *Econometric models and methods*. New York: Wiley, 1966.

Coleman, J. S. *Introduction to mathematical sociology*. New York: Free Press, 1964.

Costner, H. L., & Schoenberg, R. Diagnosing indicator ills in multiple indicator models. In A. S. Goldberger and O. D. Duncan (Eds.), *Structural equation models in the social sciences.* New York: Seminar Press, 1973. Pp. 168–199.

Duncan, B., & Duncan, O. D. Minorities and the process of stratification. *American Sociological Review,* 1968, **33,** 356–364.

Duncan, O. D. Inheritance of poverty or inheritance of race? In D. P. Moynihan (Ed.), *On understanding poverty.* New York: Basic Books, 1969. Pp. 85–110.

Duncan, O. D. Partials, partitions, and paths. In E. F. Borgatta and G. W. Bohrnstedt (Eds.), *Sociological methodology, 1970.* San Francisco: Jossey-Bass, 1970. Pp. 38–47.

Duncan, O. D. Unmeasured variables in linear models for panel analysis. In H. L. Costner (Ed.), *Sociological methodology, 1972.* San Francisco: Jossey-Bass, 1972. Pp. 36–82.

Duncan, O. D., Featherman, D. L., & Duncan, B. *Socioeconomic background and achievement.* New York: Seminar Press, 1972.

Fisher, F. M. *The identification problem in econometrics.* New York: McGraw-Hill, 1966.

Fisher, R. A. *Statistical methods for research workers* (10th ed.). Edinburgh: Oliver and Boyd, 1946.

Goldberger, A. S. *Econometric theory.* New York: Wiley, 1964.

Goldberger, A. S. On Boudon's method of linear causal analysis. *American Sociological Review,* 1970, **35,** 97–101.

Goldberger, A. S. Structural equation methods in the social sciences. *Econometrica,* 1972, **40,** 979–1001. (a)

Goldberger, A. S. Review of Blalock (1971). *Journal of the American Statistical Association,* 1972, **67,** 951–952. (b)

Goldberger, A. S., & Duncan, O. D. (Eds.) *Structural equation models in the social sciences.* New York: Seminar Press, 1973.

Goodman, L. A. A general model for the analysis of surveys. *American Journal of Sociology,* 1972, **77,** 1035–1086.

Goodman, L. A. The analysis of systems of qualitative variables when some of the variables are unobservable. Part I—A modified latent structure approach. *American Journal of Sociology,* 1974, **79,** 1179–1259.

Harman, H. H. *Modern factor analysis* (2nd ed.). Chicago: Univ. of Chicago Press, 1967.

Hauser, R. M., & Goldberger, A. S. The treatment of unobservable variables in path analysis. In H. L. Costner (Ed.), *Sociological methodology, 1971*. San Francisco: Jossey-Bass, 1971. Pp. 81–117.

Hays, W. L. *Statistics for psychologists*. New York: Holt, 1963.

Henry, N. W., & Hummon, N. P. An example of estimation procedures in a nonrecursive system. *American Sociological Review*, 1971, **36**, 1099–1102.

Jencks, C., *et al. Inequality*. New York: Basic Books, 1972.

Johnston, J. J. *Econometric methods* (2nd ed.), New York: McGraw-Hill, 1972.

Jöreskog, K. A general method for the analysis of covariance structures. *Biometrika*, 1970, **57**, 239–251.

Jöreskog, K. A general method for estimating a linear structural equation system. In A. S. Goldberger and O. D. Duncan (Eds.), *Structural equation models in the social sciences*. New York: Seminar Press, 1973. Pp. 85–112.

Keyfitz, N. Models. *Demography*, 1971, **8**, 571–580.

Klein, L. R. *An introduction to econometrics*. Englewood Cliffs, New Jersey: Prentice-Hall, 1962.

Kmenta, J. *Elements of econometrics*. New York: Macmillan, 1971.

Koopmans, T. C. Identification problems in economic model construction. *Econometrica*, 1949, **17**, 125–143. [Reprinted in Blalock (1971).]

Lazarsfeld, P. F., Pasanella, A. K., & Rosenberg, M. (Eds.) *Continuities in the language of social research*. New York: Free Press, 1972.

Li, C. C. The concept of path coefficient and its impact on population genetics. *Biometrics*, 1956, **12**, 190–210.

Li, C. C. Fisher, Wright, and path coefficients. *Biometrics*, 1968, **24**, 471–483.

Malinvaud, E. *Statistical methods of econometrics* (2nd revised ed.), Amsterdam: North-Holland, 1970.

Rao, P., & Miller, R. L. *Applied econometrics*. Belmont, California: Wadsworth Publishing Co., 1971.

Rasch, G. An individualistic approach to item analysis. In P. F. Lazarsfeld and N. W. Henry (Eds.), *Readings in mathematical social science*, Section 2. Cambridge, Massachusetts: M.I.T. Press, 1968. Pp. 89–107.

Simon, H. A. Spurious correlation: a causal interpretation. *Journal of the American Statistical Association*, 1954, **49**, 467–479. [Reprinted in Blalock (1971).]

Snedecor, G. W., & Cochran, W. G. *Statistical methods* (6th ed.). Ames, Iowa: Iowa State Univ. Press, 1967.

Stevens, S. S. Ratio scales, partition scales, and confusion scales. In H. Gulliksen and S. Messick (Eds.), *Psychological scaling: Theory and applications*. New York: Wiley, 1960. Pp. 49–66.

Stone, R. *Demographic accounting and model-building*. Paris: OECD, 1971.

Theil, H. *Principles of econometrics*. New York: Wiley, 1971.

Walker, H. M., & Lev, J. *Statistical inference*. New York: Henry Holt, 1953.

Wallis, K. F. *Introductory econometrics* (Rev. ed.). London: Gray-Mills, 1973.

Ward, J. H. Jr. Partitioning of variance and contribution or importance of a variable: A visit to a graduate seminar. *American Educational Research Journal*, 1969, **6**, 467–474.

Werts, C. E., Jöreskog, K. G., and Linn, R. L. Identification and estimation in path analysis with unmeasured variables. *American Journal of Sociology*, 1973, **78**, 1469–1484.

Wiley, D. E. The identification problem for structural equation models with unmeasured variables. In A. S. Goldberger and O. D. Duncan (Eds.), *Structural equation models in the social sciences*. New York: Seminar Press, 1973. Chapter 4. Pp. 69–83.

Wiley, D. E., & Wiley, J. A. The estimation of measurement error in panel data. *American Sociological Review*, 1970, **35**, 112–117. [Reprinted in Blalock (1971).]

Woelfel, J., & Haller, A. O. Reply to Land, Henry and Hummon. *American Sociological Review*, 1971, **66**, 1102–1103.

Wonnacott, R. J., & Wonnacott, T. H. *Econometrics*. New York: Wiley, 1970.

Wright, S. Correlation and causation. *Journal of Agricultural Research,* 1921, **20**, 557–585.

Wright, S. Path coefficients and path regressions: Alternative or complementary concepts? *Biometrics,* 1960, **16**, 189–202.

Wright, S. *Evolution and the genetics of population,* Vol. 1. *Genetic and biometric foundations.* Chicago: Univ. of Chicago Press, 1968.

Author Index

B

Blalock, H. M., Jr., 17, 24, 47, 48, 50, 90, 99, 127, 133, 150, 169
Bohrnstedt, G. W., 169
Borgatta, E. F., 169
Boudon, R., 162, 169

C

Cain, G. G., 66, 169
Carlsson, G., 168, 169
Christ, C. F., 83, 98, 169
Cochran, W. G., 8, 171
Coleman, J. S., 162, 169
Costner, H. L., 133, 169

D

Duncan, B., 50, 161, 169
Duncan, O. D., 23, 50, 66, 99, 114, 138, 147, 150, 161, 169

F

Featherman, D. L., 50, 169
Fisher, F. M., 90, 133, 169
Fisher, R. A., 50, 170

G

Goldberger, A. S., 10, 46, 90, 98, 114, 147, 150, 170
Goodman, L. A., 162, 170

Subject Index

A

Attenuation, 129
Autonomy, 151, 153, 159

B

Bias, 77, 92, 104, 107, 110,
    116–117, 120–122, 126, 138
Block-recursive, 85, 132

C

Causal inference, 20, 22, 47, 87
Causal language, 162
Causal ordering, 20, 26–28, 50,
    66, 89, 102
Cause, 1, 2, 9, 11, 15, 16
    common, 32–35, 40–42
    correlated, 37–39, 42, 75–76

Coefficient, 2, 151
Coefficient of determination, 64,
    see also Correlation,
    multiple
Cohort, 57
Correlation, 9, 10, 21, 23, 36, 50,
    127
    multiple, 63–66, 166
    partial, 15, 22–24, 48, 50
Counting rule, 83
Covariance, 6, 10, 51–52, 54–56,
    59, 68, 78, 94, 116, 126, 157

D

Dependent variable, 7, 11, 26–27,
    63, 91, 159, *see also*
    Endogenous variable
Determinant, 12, 16, 18, 30, 46
Dichotomy, 160–161

A 5
B 6
C 7
D 8
E 9
F 0
G 1
H 2
I 3
J 4